5/02

THE HUMAN GENOME PROJECT

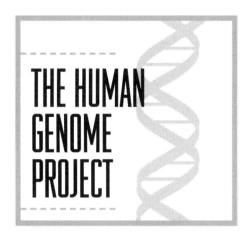

THE HUMAN GENOME PROJECT

CRACKING THE CODE WITHIN US

ELIZABETH L. MARSHALL

An Impact Book

Franklin Watts
A Division of Grolier Publishing
New York • London • Hong Kong • Sydney •
Danbury, Connecticut

For Jeff, Abigail, and Amanda

Dedicated to my colleagues from *The Scientist*

PHOTO CREDITS ©: Custom Medical Stock Photo: pp. 15, 36; Courtesy of Dr. Joan Overhauser: p. 13; Gamma-Liaison: pp. 30 (John Chiasson), 94 (Christian Vioujard), 96 (Terry Ashe); Los Alamos National Laboratory: p. 56; National Center for Human Genome Research, National Institute of Health: pp. 16, 75; Photo Researchers: p. 65 (Robert Noonan); Reuters/Bettmann: p. 49; Courtesy of Robert L. Sinsheimer, Chancellor Emeritus, U. C. Santa Cruz: p. 90; The Children's Hospital of Philadelphia: p. 59; Tommy Leonardi Photography: p. 70; Courtesy of Victor A. McKusick, M.D.: p. 85.

Illustrations created by George Stewart.

Library of Congress Cataloging-in-Publication Data

Marshall, Elizabeth L.
    The Human Genome Project: cracking the code within us / by Elizabeth L. Marshall
    p.   cm.
Includes bibliographical references and index.
Summary: Describes the fifteen-year, multimillion-dollar Human Genome Project and its process of gene mapping, includes concerns of critics of the project.
ISBN 0-531-11299-3
1. Human Genome Project.   [1. Human Genome Project.   2. Genomes.
3. Gene mapping.   4. Genetics.]   I. Title.
QH445.2.M37   1996
547.87´3282—dc20   95-45975
                                                                    CIP
                                                                    AC

CONTENTS

INTRODUCTION: WHO ARE YOU?

Who are you? Such a simple question can have many different answers. To your parents, you're the middle daughter. To your gymnastics coach, you're the one with talent on the uneven bars. To your friends, you're the funniest person in the class. Even if all these descriptions are accurate, they do not really answer the question completely.

Descriptions of age, gender, athletic ability, and personality don't tell the whole story about anybody. Adding details about your physical appearance, religion, and future plans might help round out the picture, but lots of information would still be missing.

Where you live and when you were born also influence who you are. If you had been born in another country, like Australia, would you still be you? What if you had lived 100 years ago? Or 100 years from now? Who are you?

There's yet another answer to this question. It comes from people trained in *molecu-*

lar biology. These scientists can examine a cell from your body and say something about who you are. They can characterize you by examining the forty-six *chromosomes* that are present inside each cell of your body. Within your chromosomes are *genes*, which are made of *deoxyribonucleic acid—DNA*. DNA contains information about your *heredity*, the features that you have inherited from previous generations.

By performing laboratory tests on your DNA, scientists can find out all sorts of information about you. For example, these tests can be used to determine whether you suffer from an inherited disease, such as *cystic fibrosis* or *sickle cell anemia*. They can also tell whether you are a *healthy carrier* of such diseases. To the trained eye, much of who you are is written on your DNA.

Right now, this genetic portrait of who you are is as limited as the description your gymnastics coach might give. It can describe the absence or presence of several hundred diseases or traits, and that's about all. But before too long, further study of the *human genome*, all of the genes on human chromosomes, will reveal much more.

Scientists will be able to identify the genetic errors responsible for more than 4,000 *inherited diseases*. Once scientists know what causes these diseases, they can develop ways to treat them. They will also learn more about how normal cells are programmed to grow and divide. This knowlege will increase their understanding of cancer, which is caused by uncontrolled cell division. Scientists may even discover how genes influence memory, intelligence, and behavior. In the future, the genetic description of who you are might be extremely detailed, touching upon many aspects of your identity.

Research dedicated to cracking the human *genetic code* is already well under way in laboratories around the world. This large, focused enterprise, known as the *Human Genome Project*, is dedicated to creating the tools that will allow for quick and easy exploration of the human chromosomes. Support for the international effort has come from scientists, governments all over the world, and ordinary people who are eager to learn more about human health and disease. In the United States, about $944 million had been committed to the research effort as of late 1995.

While the project promises to provide volumes of information that will lead to better treatments of genetic disease, it may also force us to face some difficult questions. What is the use of a genetic test that can predict a serious disease if no cure exists for the disease? Is it acceptable to abort a fetus with a known genetic disease? Should genetic tests be done on children?

The goal of the Human Genome Project is to create a very detailed map of the human chromosomes. This map will show the location of specific genes. Since humans have 50,000 to 100,000 genes, developing a map with enough useful detail is a major endeavor. The final map will not only show the location of genes, but will go one step further and describe each of the genes in great detail.

Completion of the Human Genome Project is scheduled for the year 2005. We don't have to wait until then for important breakthroughs, however. Many significant discoveries will occur in the process of creating the map. Early maps of chromosome 17, for example, have already helped gene hunters find the gene responsible for a specific type of breast cancer.

Yet in many ways, the end of the Human Genome Project will indeed mark a new beginning. It will provide medical researchers and biologists with a very valuable set of tools as they strive to better understand human biology.

Scientists will no doubt use these research tools in a variety of ways. After all, knowing the precise code of our genetic inheritance is one thing. Knowing how that code guides our development from a single fertilized egg and keeps our bodies in good working order is something else entirely. In many respects, mapping and sequencing the genome will raise as many questions as it answers.

Here's an analogy, one that is frequently used in connection with the Human Genome Project. In 1961, President John F. Kennedy declared that the United States would put a man on the moon by the end of the decade. For the next 8 years, hundreds of people dedicated themselves to that goal. Millions of dollars were spent designing computers, rocket engines, and other new technology. People were trained to use the new equipment. At last, on July 20, 1969, Neil Armstrong stepped upon the surface of the moon and a new era in space exploration began.

Creating a map of the chromosomes is like sending astronauts to the moon. It is an immense accomplishment. Hundreds of talented people and millions of dollars are needed. A leader who can coordinate all those people and dollars must stand at its head to make sure the job is done. But reaching the goal is not the end, just as landing on the moon was not the end of space travel. The conclusion of the Human Genome Project, the attainment of its goals, will mark the opening of a new chapter in scientific and medical research.

Not everyone is comfortable with this promise, however. From the very beginning, the Human Genome Project has faced debate and dissension. One of the most compelling concerns is that the knowledge gained will be misused. Healthy people who know they are very likely to develop an incurable genetic disease may suffer extreme depression. If parents know that one of their children will develop a genetic disease, they may treat that child differently. Couples may be pressured to use prenatal genetic tests to ensure healthy children. To hear the naysayers talk, a terrifying world of genetic and reproductive control, like the one Aldous Huxley

10

describes in his science fiction novel *Brave New World*, is close at hand.

Without a doubt, the tools developed by the Human Genome Project will be used to change our future. Both its critics and its enthusiasts agree on that. So, who is working to see that the predictions of *Brave New World* do not come true? Who is making sure that the project will be used for the greatest good?

What exactly is the Human Genome Project, and who are the people behind it?

CHAPER ONE

TINY GENETIC ERRORS

In the heart of Philadelphia, just a few blocks from the Liberty Bell and Independence Hall, are the laboratories and classrooms of Thomas Jefferson University. For the most part, the school and its hospital blend in with the bustle of street life and the classic design of the city's historical buildings. Even the Bluemle Life Sciences Building, with its brand-new brick exterior and postmodern courtyard, is easy to overlook. Yet this is where the school's biomedical research is conducted.

In a laboratory on the second floor, Joan M. Overhauser, a molecular biologist, is working on the Human Genome Project. She and her team of graduate students and postdoctoral fellows are busy studying chromosome 18.

Here is where the Human Genome Project is being carried out: in Philadelphia, Seattle, and New York; in Denver, Houston, Detroit, Salt Lake City, St. Louis, and Miami; in Newark, New Jersey, and Iowa City. Many

Molecular biologist Joan Overhauser (top right) and her team are studying abnormalities in chromosome 18. People who are missing a tiny part of chromosome 18 are mentally retarded, deaf, and have distinctive faces.

small groups of scientists, mostly at universities, are each working on one piece of the Human Genome Project. Some groups are working on specific chromosomes. Other groups are developing technologies that will make exploration of the entire genome easier, faster, and cheaper. Together, these teams are moving the project toward its ultimate goal of locating and identifying all the human genes.

In addition to the many small groups of researchers—perhaps 300 or 400 teams in all—several dozen multidisciplinary research centers are also dedicated to this effort. Among these are three national laboratories. Such large, goal-oriented research centers are something of a novelty in biological research.

Overhauser's lab and her small team of six scientists is far more typical of the way biomedical research has traditionally been conducted. A close look at her work will provide a snapshot of one small, representative aspect of the total Human Genome Project.

ABNORMAL CHROMOSOMES

Overhauser is developing tools with which to explore chromosome 18. Her interest in this particular chromosome began when she learned about a select group of people who were born with a syndrome that left them mentally retarded, deaf, and with an abnormal facial appearance. These immense problems occurred because a tiny chunk of DNA was missing from one end of chromosome 18.

Chromosomes are present in pairs, as you may know. Patients with chromosome 18 deletion syndrome have one normal copy of chromosome 18 and one abnormal copy of chromosome 18. The missing DNA means the difference between normal intelligence and mental retardation, normal hearing and deafness, a normal life and a life limited in many respects.

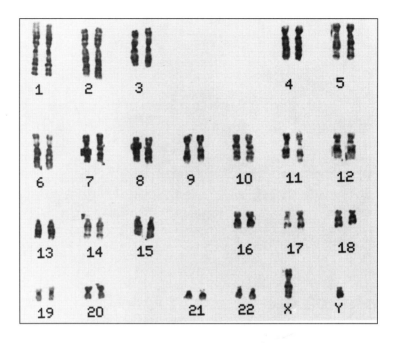

*Humans have twenty-three pairs of chromosomes. This set of
sorted chromosomes—called a karyotype—shows twenty-two
pairs plus the sex chromosomes for a male (XY). Females
have two X chromosomes (XX).*

Overhauser wants to find out what piece of DNA,
exactly, is missing from chromosome 18 in people with
this syndrome. What genes are absent? Why does the
absence of these genes cause this particular set of symptoms? If scientists knew which genes were missing, could
they somehow make up for this absence and, thus, treat
the condition?

This might be a good place to stop and review some
key concepts. The nucleus of every human cell (except
red blood cells) contains twenty-three pairs of chromosomes—that's forty-six chromosomes in all. Both of the
chromosomes in a pair look about the same. One pair of

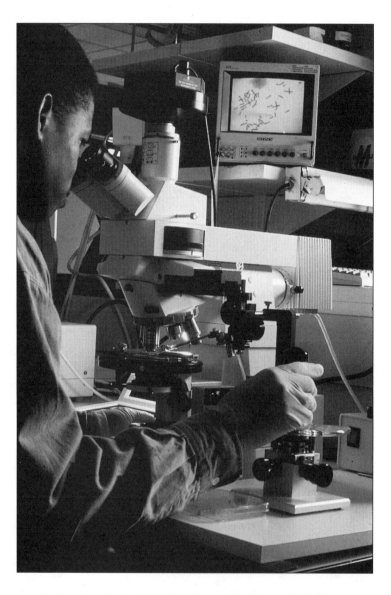

A scientist uses a powerful microscope with a video monitor to examine chromosomes. Although genes are too small to see, chromosomes can be identified by their size, shape, and pattern of bands.

chromosomes, the *sex chromosomes*, is special. If you are a female, both of your sex chromosomes look the same (XX). If you are a male, you have one X sex chromosome and one Y sex chromosome.

By exposing white blood cells to a series of different chemicals and looking at these cells under a microscope, scientists can see a person's chromosomes. It is possible to identify each chromosome by its shape, size, and pattern of bands. Scientists use these physical characteristics to arrange the chromosomes in pairs. A complete set of the sorted and labeled pairs is called a *karyotype*.

Chromosomes are made up of DNA and proteins. DNA is the molecule that carries your genetic heritage. A gene is a segment of DNA. It is too small to see, even with a microscope.

In 1953, James Watson (future director of the Human Genome Project) and Francis Crick discovered the structure of DNA. The molecule contains four building blocks called *nucleotides*: *guanine (G)*, *cytosine (C)*, *thymine (T)*, and *adenine (A)*. One strand of DNA consists of an unbroken string of these four nucleotides supported by a backbone made of sugar and phosphate groups.

The backbone consists of two strands that are twisted together in a structure called a *double helix*. A double helix looks like a spiral staircase. The molecule's two strands fit together in a very specific way, as we will see later.

THE GENETIC CODE

Your genetic code is determined by the order, or sequence, of nucleotides in a strand of DNA. Cells read this sequence as a series of three-lettered words. Each triplet of nucleotides contains instructions for building a particular *amino acid*. Scientists say that each triplet "codes" for a particular amino acid. Amino acids are linked together to build *proteins*.

For example, TATGGTGTTTCC would be read as

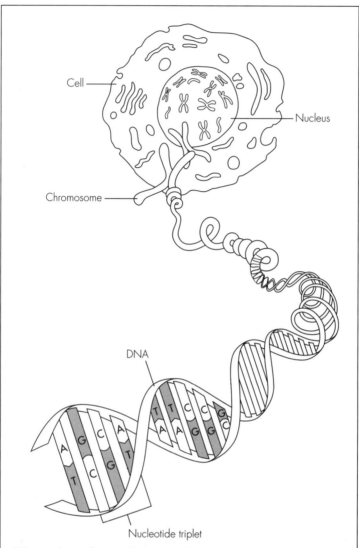

Cell

Nucleus

Chromosome

DNA

Nucleotide triplet

The nucleus of every human cell contains twenty-three pairs of chromosomes. A chromosome contains long chains of DNA nucleotides. There are four different types of nucleotide bases: guanine (G), cytosine (C), thymine (T), and adenine (A). A set of three nucleotides is a triplet.

	T	C	A	G	
T	Phenylalanine	Serine	Tyrosine	Cysteine	T
	Phenylalanine	Serine	Tyrosine	Cysteine	C
	Leucine	Serine	STOP	STOP	A
	Leucine	Serine	STOP	Tryptophan	G
C	Leucine	Proline	Histidine	Arginine	T
	Leucine	Proline	Histidine	Arginine	C
	Leucine	Proline	Glutamine	Arginine	A
	Leucine	Proline	Glutamine	Arginine	G
A	Isoleucine	Threonine	Asparagine	Serine	T
	Isoleucine	Threonine	Asparagine	Serine	C
	Isoleucine	Threonine	Lysine	Arginine	G
	START	Threonine	Lysine	Arginine	A
G	Valine	Alanine	Aspartic acid	Glycine	T
	Valine	Alanine	Aspartic acid	Glycine	C
	Valine	Alanine	Glutamic acid	Glycine	G
	Valine	Alanine	Glutamic acid	Glycine	A

Each nucleotide triplet codes for a specific amino acid. Some triplets signal a stop in the sequence. The same genetic code is used by all organisms.

four triplets: TAT GGT GTT TCC. The cell reads TAT as a command to obtain the amino acid tyrosine. GGT codes for glycine, GTT codes for valine, and TCC codes for serine. Therefore, the resulting protein would include tyrosine, glycine, valine, and serine.

Although sixty-four different triplets are possible with four nucleotides, only twenty kinds of amino acids exist. (Many triplets code for the same amino acids.) DNA instructions for various combinations and numbers of amino acids can create more than 50,000 different kinds of proteins. Some proteins are very long, with many amino acids. Others are short, with fewer amino acids.

So where do genes fit into all of this? The term gene refers to the sequence of nucleotides that codes for one protein. A gene is like a blueprint. It contains all of the information that your cell needs to build a specific protein.

Proteins are molecules that keep the body running smoothly. Cells are made of proteins, among other things. Proteins known as *enzymes* control the chemical reactions of cells. Proteins called *hormones* regulate growth, reproduction, and other biological functions.

Because genes contain the instructions needed to make proteins, they determine the color of your eyes and hair, the shape of your nose, and how tall you will grow. Some traits, like blood type, are determined by one gene. Huntington's disease, a genetic disease, is caused by one defective gene. Most frequently, however, many genes work together to determine traits. Eye color, for instance, is the result of several genes. Complex traits such as intelligence are the combined effect of many genes.

The important thing to remember is that a defective gene will build a defective protein. And a defective gene is caused by a defect somewhere in its DNA sequence. It's like the old nursery rhyme:

For want of a nail, the shoe was lost.
For want of a shoe, the horse was lost.
For want of a horse, the battle was lost.
For want of a battle, the kingdom was lost.

Except in this case, the rhyme would have to go:

For want of a nucleotide triplet, an amino acid
was lost.
For want of an amino acid, the protein was lost
(or abnormal).
For want of a protein, the body was lost (or unhealthy).

A person suffering from chromosome 18 deletion syndrome is missing a tiny chunk of DNA from one end of chromosome 18. Even though this chunk is small, it contains many genes. As a result, many proteins do not form correctly. Because these proteins cannot do their jobs, the person has the symptoms of chromosome 18 deletion syndrome. Even though the person has one normal copy of chromosome 18, that's clearly not enough to compensate for the portion that is missing from the other chromosome.

Yet having too many copies of chromosome 18 has even larger consequences. Babies born with three copies of chromosome 18 suffer from Edwards syndrome (also called trisomy 18), a condition marked by profound mental retardation and severe heart disease. More than 90 percent of these patients die before their first birthday.

"You're probably never going to be able to cure chromosomal syndromes," Overhauser says. "But you can try to figure them out. You can try to understand something about what's going on and something about normal brain development."

On this spring day, Joan Overhauser is dressed in casual pants and a patterned pink-and-purple vest. Large triangular earrings dangle from her ears. If she were to stand beside the graduate students and postdocs on her team, it might be difficult to pick her out as the project leader. Yet when she speaks about her work she is clear, careful, and earnest. It becomes easy, then, to imagine her lecturing to a class of medical students or critiquing the lab's work.

"I got into this field because it was very important to me to have some kind of an impact," Overhauser adds. "Everybody is motivated in different ways. I felt that for me to keep motivated I wanted my work to have some kind of medical relevance."

The job Overhauser has set for herself contains several goals. First, she wants to locate the genes that are missing in people with chromosome 18 deletion syndrome. After she isolates and identifies the genes, she can then learn what proteins they encode. Finally, once she knows the identity of the missing proteins, she can begin to think about how the absence of these proteins manages to disrupt normal development and cause mental retardation, deafness, and the other symptoms.

In 1988, when Overhauser was a young scientist newly hired by Thomas Jefferson University, her goals and the goals of the Human Genome Project happened to coincide. That was the year she applied to the National Institutes of Health (NIH), a government agency, for the money to start her lab. The NIH directors decided to fund her through the brand-new Human Genome Project. They realized that Overhauser's work with chromosome 18 dovetailed nicely with one of the Human Genome Project's most important goals: mapping the human chromosomes.

MAKING MAPS

Where does a scientist like Overhauser begin? How does she find missing genes? How does she keep track of the genes that she finds? Such simple-sounding questions are really at the heart of the Human Genome Project. Unfortunately, there are no simple answers. Working with DNA, despite the spectacular advances in molecular biology of the last decade or so, is still quite difficult. Scientists use a variety of approaches and tools.

The directors of the Human Genome Project welcome such diversity among the scientists they fund. They know that, as scientists share research results, successful strategies will spread and flourish and less effective techniques will lose popularity.

Working with DNA is challenging because it is very small and very long. An enormous number of nucleotide base pairs are scrunched into a very tiny space. Even though chromosome 18 is shorter than most of the other chromosomes, its DNA consists of about 90 million nucleotide base pairs.

Overhauser estimates that 1,500 to 3,000 genes are present on chromosome 18. This estimate is imprecise because there is no easy way to count genes. You can't see genes, after all. Although an extremely powerful microscope, known as a scanning, tunneling microscope, can show a close-up picture of DNA, this image can't tell you the sequence of nucleotides. It can't even tell you if you're looking at a gene. A gene, remember, is a section of DNA that codes for a protein. Only by using various laboratory techniques can scientists deduce whether a section of DNA is a gene.

Exploring DNA, to use a rough analogy, is rather like drawing a map of a room you're not allowed to enter based upon whom you see entering and exiting. It's like studying animal tracks in the forest and concluding that bears are present.

Or here's another example. Think of two bottles— one contains water and the other contains ammonia. The two liquids look alike. You can't distinguish them by sight. Instead, you might try experimenting with them to determine their properties. Maybe you'd sniff for fumes. If you were very lucky, you might find that someone else had solved the problem before you and had labeled the bottles.

Faced with an immense, uncharted territory like DNA, scientists do what explorers have always done: They draw pictures. They label. They locate landmarks. They figure out the order of things. They make maps.

Scientists like Overhauser talk about "mapping the genome." But "map" can be a confusing word if you think of a map as a flat piece of paper. A poster of a human skeleton shows all the bones and their locations relative to each other. You might call it a map of the skeletal system. But an actual skeleton, hanging in a doc-

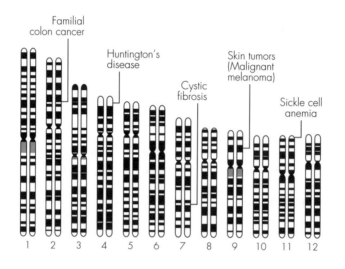

Familial
colon cancer

Huntington's
disease

Cystic
fibrosis

Skin tumors
(Malignant
melanoma)

Sickle cell
anemia

1 2 3 4 5 6 7 8 9 10 11 12

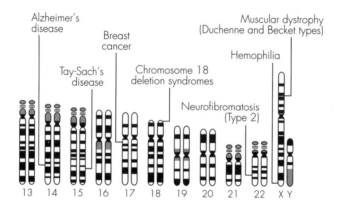

Alzheimer's
disease

Breast
cancer

Tay-Sach's
disease

Chromosome 18
deletion syndromes

Muscular dystrophy
(Duchenne and Becket types)

Hemophilia

Neurofibromatosis
(Type 2)

13 14 15 16 17 18 19 20 21 22 X Y

This physical map of the human genome shows the locations of many disease genes. Staining a cell creates a distinctive banding pattern on its chromosomes. Scientists can use these bands as landmarks to describe the locations of specific genes.

tor's office, does the same thing. In a broad sense, isn't that a map too?

Keep these "maps" in mind as you take a look at the work of Overhauser and other scientists. Their goal is to create a way to describe the locations of specific sections of DNA. That's what it means to map the genome. Researchers want answers to two related questions: Where on the chromosome is this gene? Where is this gene in relation to other known genes? To answer these questions, scientists create different kinds of maps.

One type of map created by genome scientists is called a *physical map*. Many types of physical maps exist. A picture of a chromosome stained to show its distinctive bands is one kind of physical map. A scientist can describe the location of a gene by stating that it is on a particular band. This helps narrow the location of a gene, although many genes are on each band. A list of genes in the correct order is also a useful physical map. The most detailed physical maps describe the distance, in nucleotides, between genes.

One of the goals of the Human Genome Project is to divide the entire human genome into relatively short sections of DNA, place each DNA segment in its own laboratory dish, and label the dishes. Such a collection of dishes, arranged in the same order as the DNA segments actually appear on the chromosomes, is considered to be the ultimate physical map. This kind of physical map is also known as a DNA library. Just as you find an entry in the encyclopedia by first selecting the proper volume, scientists who wish to study a particular gene can select the proper dish of DNA from their copy of a DNA library.

Working with chromosomes is not easy. To create physical maps, genome scientists must first find innovative ways to approach DNA.

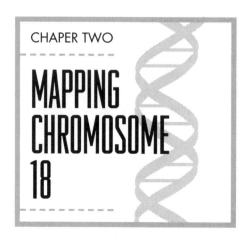

CHAPER TWO

MAPPING CHROMOSOME 18

Chromosomal deletions offer clever scientists like Overhauser a starting point for their explorations. Rather than contemplate an entire chromsome, a deletion allows Overhauser to concentrate on one relatively small area of a chromosome. In people with chromosome 18 deletion syndrome, the DNA segment that is missing is of a useful size, big but not too big. Although the size and location of the deletion varies from person to person, scientists estimate a typical deletion to be about 3 million base pairs long and to contain 50 to 100 genes. This is a workable size for the laboratory techniques Overhauser uses.

Studying chromosomal deletions means studying people, of course, and all of Overhauser's DNA experiments begin when she reaches out to a family with a child who is suffering from a deletion syndrome. Understandably, some parents are skeptical, so Overhauser has made an effort, over the years, to introduce herself to parent support groups and to give short speeches about her genetics research. She carefully explains how she hopes to discover the genes that are missing from the one chromosome 18 with the deletion.

Still, it is a commendable leap of faith for a family to give Overhauser permission to study their child's medical history and to draw blood samples. Overhauser's work will quite possibly improve the lives of future patients. It may even, one distant day, lead to a cure for the syndrome. But it is extremely unlikely that anyone with a deletion syndrome today will live to see those results.

Once Overhauser obtains blood from people with the deletion syndrome, she knows she has the faulty chromosome in her possession. It is there, in every white blood cell of her sample, but not in a form that is useful for study in the laboratory. The molecular biologist needs to separate chromosome 18 out of the blood cells. Indeed, this is a major challenge of DNA science: How can something as tiny and fragile as a chromosome or a section of DNA be isolated and kept whole? How can it be studied?

STUDYING DNA

Overhauser's solution is to combine her sample of human cells with hamster cells. She then treats these hybrid cells in a special way so that only one copy of chromosome 18—the copy with the deletion—remains in the cell. With this vehicle—the hamster-cell hybrid—Overhauser propagates the chromosome. To examine the chromosome more easily, she uses other techniques to make many copies of the sample. She can then begin designing experiments, knowing that her inventory of chromosome 18 is in good shape.

Now Overhauser is ready to figure out which sections of DNA are missing from the abnormal chromosome. To do that, she makes use of an important property of DNA molecules. As mentioned before, DNA consists of two strands that are twisted together in a double helix. For this reason, DNA is described as double stranded.

Its two strands of nucleotides match up according to two simple rules: adenine (A) pairs with thymine (T), and cytosine (C) pairs with guanine (G). For example, a stretch on one strand of DNA that reads ACGTAT can only match with its complementary sequence TGCATA on the other strand. Thus, to know the code on one strand is also to know the code on the other. If the DNA molecule is stripped of one of its strands, that strand will seek out its complementary partner.

Overhauser separates the two strands of DNA from a normal chromosome 18 and exposes it to an enzyme that cuts it into hundreds of fragments. She doesn't know much about these fragments—their DNA sequence, their location on the chromosome—but she does know that they all come from chromosome 18. That's all she needs to know for the moment. She is now able use these anonymous fragments as *probes* with which to explore the abnormal chromosome of patients with chromosome 18 deletion syndrome.

By adding the anonymous fragments of DNA to fragments from the abnormal chromosome 18, Overhauser can observe which probes find a partner and which don't. Most probes will find their complements and stick to the abnormal chromosome. But a few probes will fail to find their partner. Those probes, those lonely, unmatched probes, become particularly important. Their failure to find a match indicates that their partner is missing. In other words, their partner is on the deleted portion of the chromosome and the absence of those genes is causing the symptoms of the chromosomal syndrome.

This is an oversimplification of the process, of course. In the laboratory, Overhauser or her associates must go through a series of careful steps that include inserting the DNA fragments from the abnormal chro-

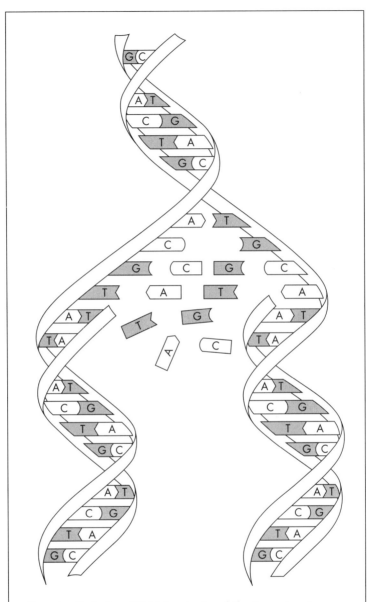

In a double helix of DNA, adenine (A) always bonds to thymine (T) and cytosine (C) always bonds to guanine (G).

Scientists examine images of DNA fragments on a piece of photographic film. Examining patterns of DNA fragments gives researchers a way to study tiny genetic differences, like chromosomal deletions or nucleotide sequences.

mosome into a gel, putting the gel in an electric field to separate the DNA fragments by size, and transferring the fragments to a nylon membrane. The nylon membrane is exposed to the anonymous probes from the normal chromosome, which have been radioactively tagged. This makes them easy to see when they are exposed to X-ray film. The probes that find a partner on the membrane will stick and make a mark at a specific location on the film. This method of determining which DNA strands have stuck to each other is known as DNA *hybridization*.

SEARCHING FOR DELETIONS

Probe after probe, Overhauser repeats the process. She and her team look closely for differences among the hybridization patterns of the patients' chromosomes. On certain abnormal chromosomes, few probes find a partner. She can then deduce that these chromosomes contain large *deletions*. On other chromosomes, many of the probes stick. This indicates a small deletion. By comparing these results, Overhauser can make deductions about the locations of the probes.

Here's an example. Imagine that you are a molecular biologist studying the chromosomes of three patients—Anna, Tiffany, and Bob—with chromosome 18 deletion syndrome. You have four anonymous probes, labeled 2, 4, 9, and 11. You will use these to determine the sizes of your patients' deletions.

Probes 9, 11, and 4 map to Bob's chromosome, but probe 2 doesn't. He has a relatively small deletion. Only probe 11 maps to Tiffany's chromosome. She has a large deletion. Probes 11 and 4 map to Anna's chromosome. Anna's deletion is smaller than Tiffany's but larger than Bob's.

From this study you can conclude that the DNA represented by probe 11 must be farthest from the end of the chromosome, because that's the only probe that hybridized to Tiffany's chromosome. Suddenly the probes are a bit less anonymous! Probe 4 is next (probes 11 and 4 hybridized to Anna's chromosome), then probe 9 (probes 11, 4, and 9 hybridized to Bob's chromosome), and finally probe 2. Now you know the order of the probes—11, 4, 9, 2. You could begin to construct a physical map of chromosome 18.

One of Overhauser's first maps, published in the journal *Genomics* in 1992, plotted 200 probes to eleven regions on chromosome 18. Her anonymous fragments were no longer anonymous. She could label them

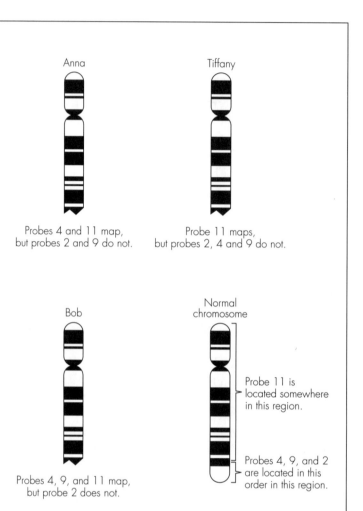

Anna

Probes 4 and 11 map,
but probes 2 and 9 do not.

Tiffany

Probe 11 maps,
but probes 2, 4 and 9 do not.

Bob

Probes 4, 9, and 11 map,
but probe 2 does not.

Normal
chromosome

Probe 11 is
located somewhere
in this region.

Probes 4, 9, and 2
are located in this
order in this region.

*Under the microscope, Tiffany, Anna, and Bob seem to have
the same deletion. But by using probes, subtle differences
can be detected. With this approach, you can conclude that
Tiffany has the largest deletion, followed by Anna and Bob.
You can also deduce the following order of the probes from
the end of the chromosome: 2, 9, 4 and 11. Number 2 maps
closest to the end.*

according to the region in which they appeared. Indeed, they could serve as landmarks on the chromosome.

Additional experiments with those fragments will ultimately result in an ordered set of DNA fragments. This, too, is one kind of physical map. Overhauser's map might eventually be used by other scientists interested in investigating a particular segment of DNA on chromosome 18.

Her map has already offered hints toward a better understanding of chromosome 18 deletion syndrome. It seems, as you may have guessed, that people with larger deletions have more symptoms, and more severe symptoms, than people with smaller deletions. The size of the deletion seems to correspond with the severity of medical problems.

For instance, some people with the deletion syndrome have normal-size heads and others have small heads. Using her map, Overhauser was able to show that the patients with the largest deletions had unusually small heads. Patients with smaller deletions had normal-size heads. This observation has prompted Overhauser to hypothesize that the location of the genes responsible for head growth are located in a region relatively far from the end of the chromosome. The deletions of most patients don't reach that far.

SHARING RESOURCES

To move this research along faster, Overhauser's lab shares resources with many laboratories around the world, including teams in the Netherlands and Germany. By sharing probes and comparing maps, the teams can help each other. The scientists have also been able to meet face to face almost every year since 1992 at an international workshop funded by the Human Genome Project and the Human Genome Organization, an international group.

Overhauser credits the Human Genome Project for this kind of collaboration. "The Human Genome Project has allowed people that normally wouldn't interact to come together bringing their different areas of expertise," she says. "You can never force anybody to collaborate. But you can create an atmosphere that helps things along."

The physical map of chromosome 18 continues to grow in detail. It still has a long way to go, however. In 1992, Overhauser was able to divide the chromosome into eleven regions of approximately 8 million nucleotide base pairs each. By 1998, the directors of the Human Genome Project would like to see the genome divided into regions no longer than 100,000 base pairs. This will require locating and identifying about 30,000 DNA landmarks along the genome.

The ultimate map—the one with the most detail—will show the exact sequence of all the nucleotide base pairs on the genome. Further technological advances are needed before such a detailed map can be created. This isn't expected to happen until the year 2005, when the Human Genome Project is scheduled for completion. In the meantime, less detailed maps can still be very useful, as Overhauser's work demonstrates.

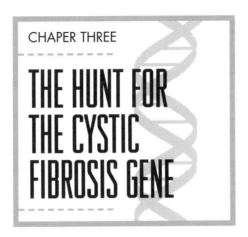

THE HUNT FOR THE CYSTIC FIBROSIS GENE

On August 24, 1989, one of the most dramatic hunts for a human gene formally came to an end. On that day, scientists from the Howard Hughes Medical Institute at the University of Michigan and the Hospital for Sick Children in Toronto announced that the cause of cystic fibrosis is a tiny deletion on chromosome 7.

For boosters of the Human Genome Project, this was an enormous triumph. Here was proof of the power and potential of the Human Genome Project. By conducting a painstaking search through the human genome, the cause of a deadly disease had been discovered. Surely a more complete exploration—like that endorsed by the Human Genome Project—could solve other tragedies of human disease.

People with cystic fibrosis produce unusually large amounts of sticky mucus. This mucus clogs the small passages of the lungs, making it very difficult to breathe. It also cre-

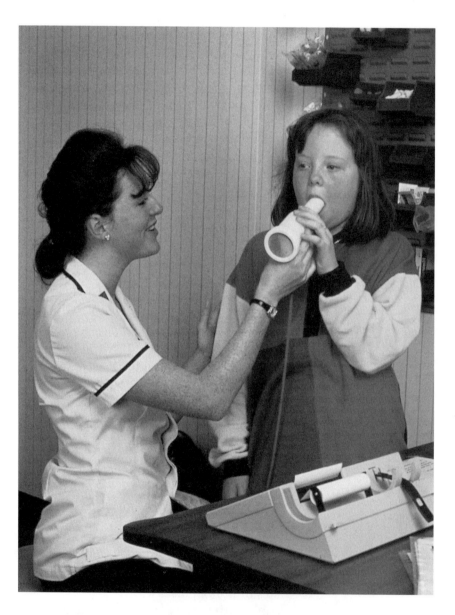

A girl with cystic fibrosis tests the capacity of her lungs. Mucus clogs the small passages of the lungs in people with cystic fibrosis, making it difficult to breathe.

ates an environment for dangerous, life-threatening infections. About 30,000 people in the United States have cystic fibrosis. The severity of the condition varies. Some patients die during childhood, while others may live into their thirties or forties.

In 1989, scientists discovered that individuals with cystic fibrosis are missing a single nucleotide triplet in the middle of both copies of chromosome 7. Further investigation showed that this deletion causes a defect in the protein responsible for moving salt and water in and out of cells lining the nasal passages and the lungs.

CYSTIC FIBROSIS

Cystic fibrosis is the most common lethal genetic disease in the United States. It usually affects people of European ancestry; African-Americans, Hispanics, and Asians rarely suffer from the disease. Until fifty years ago, children with cystic fibrosis were thought to suffer from severe pneumonia. Eventually, however, physicians and scientists realized that the disease was not caused by a virus or bacteria. Rather, it was an illness that people were born with, a cruel destiny inscribed in their genes.

By studying large families in which one or more children had the disease, scientists realized that the gene for cystic fibrosis had to be *recessive*. This means that to get the disease, a person must have two defective copies of the CF gene. People with one normal copy and one defective copy are healthy. But if two healthy carriers have a child, that child has a one in four chance of inheriting two defective copies of the gene. This explains how a healthy mother and a healthy father can have both healthy children and children with cystic fibrosis.

If a genetic disease is caused by a *dominant gene*, a single defective copy will cause the illness. With dominant genes, there is no such thing as a carrier. The genes for blood types A and B, for example, are dominant. A domi-

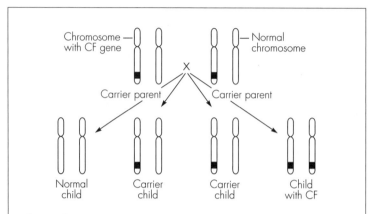

Cystic fibrosis is a recessive disease. A person must have two copies of the disease gene to be affected by the disease. A person with only one copy of the gene is unaffected. Two unaffected parents can have a child with cystic fibrosis if each is a carrier of the disease gene. Each child they conceive has a one in four chance of inheriting two disease genes.

nant gene on chromosome 4 causes Huntington's disease. Because this genetic disorder does not develop until the age of 40 or 50, individuals often do not know that they are passing the gene to their children.

The hunt for the cystic fibrosis gene began in 1982, three years before the Human Genome Project was even a dream. Indeed, the search for the cystic fibrosis gene, as it was conducted, might never have qualified for Human Genome Project funds. The gene hunters hot on the trail of the gene were not interested in mapping chromosome 7. They were not even interested in finding other genes. They had one goal, and one goal only: to identify the flawed section of DNA that was responsible for cystic fibrosis.

Scientists investigating cystic fibrosis faced a daunting task. Where should they begin? At that point, they

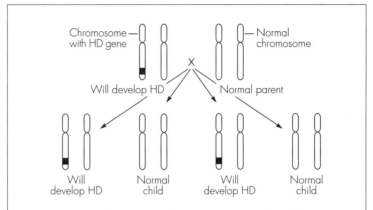

Chromosome— with HD gene

Normal chromosome

X

Will develop HD

Normal parent

Will develop HD

Normal child

Will develop HD

Normal child

Huntington's disease is a dominant genetic disorder that usually develops after age 40. It affects the nervous system and causes early death. A person needs only one copy of the gene to be afflicted by the disease. A person with a dominant disease gene has a one in two chance of conceiving an affected child.

didn't even know which chromosome the abnormal gene was on. Nothing about the disease itself suggested the best place to begin searching. They couldn't guess the identity of the genetic error from their knowledge of the deadly lung mucus. It was even possible, although unlikely, that cystic fibrosis was caused by more than one defective gene. "Looking for a cystic fibrosis gene was like looking for a house in a city between New York and Los Angeles without a street address," said Lap-Chee Tsui. Tsui (pronounced "Choy") is a scientist studying cystic fibrosis at the Hospital for Sick Children in Toronto.

So the teams started with what they did know. They knew the faces and voices of the children and young adults who suffered from the disease. The scientists also knew important genetic details about those patients and their families. They knew that those young people carried two copies of the defective gene in their cells. The researchers knew that each of the patients' parents must be "silent carriers," with one flawed gene. And the patients' healthy siblings might have either one or no flawed genes.

LAP-CHEE TSUI'S TEAM

The Hospital for Sick Children in Toronto has one of the largest cystic fibrosis clinics in the world. Hundreds of families, including some with two or more children suffering from cystic fibrosis, have gone to Toronto for help.

In 1982, Tsui began collecting blood samples from as many members of these cystic fibrosis families as possible. By studying the DNA of affected and unaffected relatives, he and his colleagues hoped to find clues that would lead them to the lethal gene. As their search progressed, Tsui made maps of the human chromosomes recording his research.

Lap-Chee Tsui's map is not like Joan Overhauser's, however. Her map places known DNA segments in the proper order and measures the actual distance, in terms of nucleotide bases, between them. These recognizable segments can then be used as *markers*, serving as sign posts along the chromosome. Tsui's map also shows the relationship between markers, but the closeness of the relationship is measured according to how often they are *linked* or inherited together. A marker maybe a gene or it may just be a distinct stretch of DNA.

To understand *genetic linkage mapping* you need to understand something about what happens when egg and sperm cells are formed. The genes on a chromosome are never separated—except during the production of egg and sperm cells. When this process (known as *meiosis*) occurs, each chromosome splits into two *sister chromatids*. Each chromosome pair then lines up side by side. Some-

The Steps of Meiosis

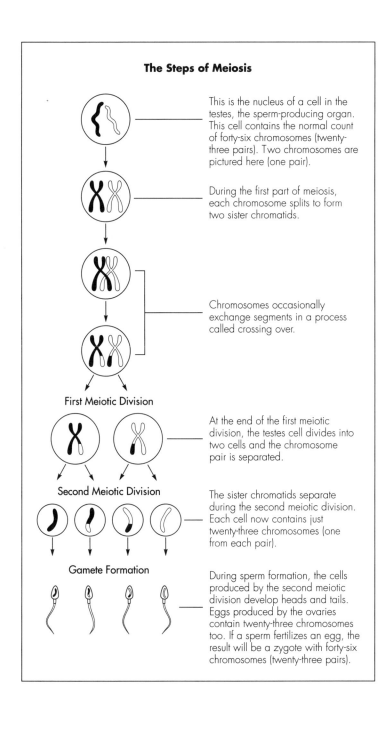

This is the nucleus of a cell in the testes, the sperm-producing organ. This cell contains the normal count of forty-six chromosomes (twenty-three pairs). Two chromosomes are pictured here (one pair).

During the first part of meiosis, each chromosome splits to form two sister chromatids.

Chromosomes occasionally exchange segments in a process called crossing over.

First Meiotic Division

At the end of the first meiotic division, the testes cell divides into two cells and the chromosome pair is separated.

Second Meiotic Division

The sister chromatids separate during the second meiotic division. Each cell now contains just twenty-three chromosomes (one from each pair).

Gamete Formation

During sperm formation, the cells produced by the second meiotic division develop heads and tails. Eggs produced by the ovaries contain twenty-three chromosomes too. If a sperm fertilizes an egg, the result will be a zygote with forty-six chromosomes (twenty-three pairs).

Two Kinds of Chromosome Maps

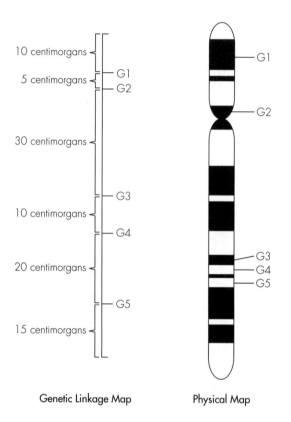

Genetic Linkage Map   Physical Map

On a physical map, the distance between genes is measured in nucleotide base pairs. On a genetic linkage map, distance is measured in centimorgans. Genetic linkage maps are based on laboratory observations. Although the order of genes is the same on both maps, the distances on the two maps will not correspond. In this diagram, the genes are labeled G1, G2, G3, G4, and G5.

times the paired chromosomes swap sections of DNA to make a new, never before seen chromosome. This swapping action is called *crossing over* or *genetic recombination*.

DNA segments that are far apart on a chromosome are often separated during crossing over. But segments that are close together will usually stay together and, thus, are inherited together generation after generation. On linkage maps, distance is measured in terms of the likelihood that two segments will be separated during crossing over.

A *centimorgan* is the unit of measurement used by these maps. (It's named in honor of geneticist Thomas Hunt Morgan.) If two DNA segments are 60 centimorgans apart, that means they have a 60 percent chance of being separated during meiosis.

GENETIC LINKAGE MAPS

Here's another way to think about genetic linkage maps. Whom do you see very often? The friends and family who live near you, of course. Whom do you see only once or twice a year? The relatives, perhaps grandparents or cousins, who live far away. By closely studying the people who enter and exit your house, observers could chart how often a certain person appears there.

These observers might deduce that the brown-haired teenager who is always with you lives nearby, while the elderly man with a cane who comes only at Christmas lives far away. The observers don't know exactly how close, in miles, the teenager lives nor how far away the old man lives, but they do know that the old man lives farther from you than the teenager.

Tsui's maps grew in detail as he and his team conducted the long, slow, tedious process of comparing the DNA from healthy relatives to the DNA of the people with the disease. Which similarities were important? Which differences were important? In this family, were

Fragment A   Fragment B

Scientists use restriction enzymes to help identify differences in the nucleotide sequences of DNA segments. One type of restriction enzyme seeks out the sequence GAATTC and cuts it between G and A.

certain distinct sections of DNA inherited along with the defective gene? Finally—and this was the crucial question—could such DNA markers lead the Toronto group to the location of the abnormal gene?

The scientists needed to find markers that showed some variability from one person to the next. Such a search is difficult because most people have very similar DNA. Oh, we each might have a unique face, but still it's a *face* rather than a snout or a flower blossom. All the DNA that we share in common makes us identifiable as human beings rather than as wart hogs or petunia plants. Out of the total 2.8 billion nucleotide base pairs in the human genome, any two people will have exactly the same sequence for 2,786,000,000 of those base pairs. Only about 14 million nucleotide base pairs (0.5 percent) will be different.

The trick, then, is to focus on the locations of the 14 million base pairs that do show variations. That comes out to one slight difference every 500 or 600 nucleotide base pairs. To locate those differences, scientists use special enzymes that seek out particular combinations of nucleotide bases. When the *restriction enzyme* finds these specific combinations, it cuts them. One such enzyme, for instance, searches for the combination GAATTC. When the enzyme finds this sequence of nucleotides, it cuts the DNA between the G and A nucleotides.

It's extremely unlikely that two unrelated people will have exactly the same number and pattern of nucleotide combinations, especially if the combination is rather large. One person might have the key combination ten times on a particular DNA segment. The enzyme would cut this segment ten times to create eleven pieces of varying sizes. Another person might have the key pattern only three times. Cutting it would result in four, relatively large pieces.

Studying the size and number of DNA fragments after cutting the DNA with a restriction enzyme shows how different or similar the DNA from two people is. Any variation in DNA is known as a *polymorphism*. (The word is Greek meaning "many forms.") Eye color in humans is a polymorphic trait caused by the combined effects of normal DNA polymorphisms. A variation in the DNA sequence that can be identified with restriction enzymes is called a *restriction fragment length polymorphism*, or RFLP. (This is pronounced "rif-lip.") RFLPs can be very ueful as markers along the chromosome.

SEARCHING FOR RFLPS

For three years, Tsui searched DNA fragments taken from all the chromosomes for a RFLP that was shared by people who were known to carry at least one copy of the defective CF gene. This included the cystic fibrosis patients (one pair of flawed chromosomes, identity unknown) and their healthy parents (one flawed chromosome).

Tsui concentrated on eleven families that contained two or more sick children. If a RFLP marker was present in people with flawed chromosomes, but absent in people with normal chromosomes, this was a very good indicator that the RFLP and the defective gene were closely linked. Knowing the location of the RFLP meant knowing the general location of the flawed gene.

Tsui's genetic linkage maps grew in detail. He shared his DNA probes and his family trees with other teams conducting the same search. Then, in 1985, Tsui and two other teams confirmed that a RFLP linked to the flawed gene had been found. It was located on chromosome 7. Each of the three teams announced their discovery in separate articles that appeared in November of that year.

An enormous breakthrough, yes, but also the beginning of another long and difficult search. Where on chromosome 7 was the abnormal gene? Where, in relation to the known RFLP marker, was it located? By studying his linkage maps, Tsui estimated that the RFLP was within 15 centimorgans of the defective gene that caused cystic fibrosis. In terms of physical distance, he estimated that it was 15 million base pairs away. What was needed were more markers that were more tightly linked to the gene.

Months passed. At last, in December 1986, a team at the University of Utah published a report in the *American Journal of Human Genetics* announcing that they had found two markers—one on either side of the suspect gene. Somewhere in a region of 1.6 million base pairs, a segment large enough to contain 50 to 100 genes, lurked the gene they sought.

The scientists couldn't hope for much better. It was time to close in on the gene itself. To carry out this mission, Tsui joined forces with a gene-hunting team led by Francis Collins. Collins was a Howard Hughes Medical

Institute investigator at the University of Michigan in Ann Arbor. This meant that the Howard Hughes Medical Institute, a private organization with an enormous amount of money and a commitment to biomedical research, funded his work.

Standard laboratory technique for closing in on a gene calls for "walking" the chromosome until the gene is found. Walking involves overlapping fragments of DNA to capture the desired sequence. Collins had improved upon that technique and created something he called "jumping." Together, Tsui's and Collins's teams jumped and walked across 280,000 nucleotide bases before encountering the beginning of the gene that causes cystic fibrosis.

How did they know this was the right gene? They didn't, at first. They began by figuring out the correct order of its nucleotide base pairs, a slow and expensive process known as "sequencing." With the exact sequence of the nucleotide triplets in hand, they could then figure out which amino acids made what protein. When they analyzed the kind of protein the gene coded for, they knew they had found the right one.

In its normal form, this gene codes for a protein that moves salt and water across the cell membranes in the airways. Presumably if the membrane protein didn't work, the lungs wouldn't work right either. Eureka!

Scientists discovered that the gene for the membrane protein was quite large, nearly 250,000 nucleotide base pairs long. Not all of those nucleotides played a role in building the protein, however. Discoveries beginning in the early 1980s had revised the notion that every nucleotide triplet in a gene must code for an amino acid. Instead, it seems that genes contain coding and noncoding regions. (What are the noncoding sections good for? No one knows for sure yet, but by the conclusion of the Human Genome Project scientists may have some answers.)

The CF gene consists of twenty-seven *exons*, DNA that codes for the protein. Between two exons, there is a noncoding region called an *intron*. The exons of the CF gene contain a total of 4,440 nucleotides that code for 1,480 amino acids.

Seventy percent of cystic fibrosis patients have a three-nucleotide-base deletion. The missing triplet should code for the amino acid phenylalanine (TTT or TTC). This deletion is near the middle of the gene, at position 508, and is thus known as the phenylalanine 508 deletion. Think of it—a missing chunk this small means the difference between good health and years filled with life-threatening emergencies. Similar tiny defects elsewhere in the gene are responsible for the other 30 percent of patients. It seems that errors anywhere in the gene result in a malfunctioning protein.

THE ANNOUNCEMENT

The teams' good news was impossible to contain. Scientists usually announce their discoveries in articles published in scholarly journals. Before publication, the journal editors make sure that the research is important and carefully done. In keeping with this tradition, scientists wait for journal publication before discussing their work with the press or the public.

These sacred traditions fell apart completely in the face of the cystic fibrosis gene discovery. News of the journal report, scheduled for publication in the September 8 issue of *Science*, leaked out. Bowing to pressure, *Science* magazine broke its own rules and permitted Collins, Tsui, and Jack Riordan (also a member of the Toronto team) to hold a press conference. Actually, they held two, one in Toronto and one in Washington, D.C., on August 24, 1989. To make this whirlwind of publicity possible, the Howard Hughes Medical Institute kindly provided private jets.

Jack Riordan (left), Francis Collins, and Lap-Chee Tsui
announced the discovery of the defective gene that causes cystic
fibrosis at a news conference held on August 24, 1989.
In 1993, Collins became the director of the Human
Genome Project.

Probably those most thrilled by the announcement were cystic fibrosis patients and their friends and family. Now that the gene had been located, new approaches to treatment could be tested. Prenatal diagnosis, of particular interest to couples who already had one child with the illness, was possible. Most exciting of all, therapies designed to substitute healthy genes for faulty ones could begin.

Gene therapy promises to cure genetic disease by giving patients what they lack: healthy genes. For cystic fibrosis sufferers, this means receiving normal genes for the membrane protein needed to keep their lungs clear of mucus.

Gene therapy experiments began at institutions around the country soon after the cystic fibrosis gene was identified. Although researchers saw some early successes with cells growing in laboratory cultures, the strategy has worked far less well with human patients. Researchers still have a great deal of work ahead of them. Nevertheless, the discovery of the cystic fibrosis gene remains a remarkable milestone in the search for a cure.

The hunt for the cystic fibrosis gene also did several important things for the Human Genome Project.

First, it raised Collins to a new level of prominence within the biomedical community. Both a research scientist and a physician, he became the Human Genome Project's second director, after James Watson resigned in 1993.

Second, it raised a question that couldn't be ignored. The gene hunt had cost $120 million, mostly taxpayer dollars spent as government research grants. Other big gene hunts promised to spend as much. Wouldn't it make good financial sense to map and sequence the entire genome once and for all?

Finally, and most significantly, it demonstrated that the tools of genomic research, especially the use of genetic linkage maps and RFLPs, really could change the face of medicine. That had always been the promise that the genome project held out. Now, with one dramatic example, the promise was a reality. In the years to come, other kinds of DNA markers would replace the use of RFLPs and new advances in genetic linkage mapping would be made. Still, no discussion of the Human Genome Project seems complete without a triumphant remark about the discovery of the cystic fibrosis gene.

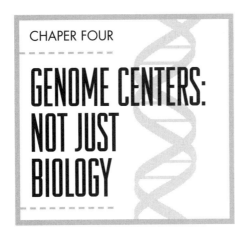

GENOME CENTERS: NOT JUST BIOLOGY

Many small groups carry out the work of the Human Genome Project. Yet small groups alone aren't enough to finish the project. To completely map and sequence the human genome, enormous advances in technology must occur. Computer systems must be developed to allow access to information gathered by the project. Innovative new tools for quickly and cheaply sequencing DNA must be created.

Biologists can't build such things by themselves; they need the help of mathematicians, chemists, engineers, computer scientists, and physicists. And very often the best place to collaborate with these kinds of people is in a large scientific center specifically dedicated to genome work.

Much of the Human Genome Project money spent by the two government agencies running the project—the Department of Energy (DOE) and the National Institutes of Health (NIH)—is concentrated in less than two dozen centers. DOE spent 60 percent of its genome budget on three centers located at national laboratories. The NIH National Center for Human Genome Research spends about 40 percent of its budget on eighteen large centers. Most of these Genome Science and Technology Centers are located at universities. A few are at private institutions or commercial companies.

NIH also sponsors genome research on its own campus in Bethesda, Maryland. Francis Collins was responsible for creating this program, known as the Division of Intramural Research within the National Center for Human Genome Research, in February 1993, shortly after he became the director of NIH's genome effort.

This program is separate from the Human Genome Project and has very different goals. Rather than focusing on developing the tools and technology for genome research, the scientists in Bethesda are dedicated to disease-gene hunting. They work with other NIH scientists and physicians who are studying human disease. This program receives about 25 percent of the NIH genome budget.

The genome centers located outside of Bethesda are like the Human Genome Project in miniature. Just as the Human Genome Project organizes and encourages a vast array of mapping and sequencing efforts in this country, the big genome centers encourage a variety of goals too. They nurture large-scale, multidisciplinary efforts that cannot be completed in any other setting.

As high-profile representatives of the genome effort, these large centers also shoulder special responsibilities. They are obligated to see that discoveries and resources are shared quickly with the scientific community. They also have the resources to offer seminars about the Human Genome Project to journalists, teachers, and the general public. Indeed, sponsoring such educational opportunities is part of their mission.

Yet even with this kind of diversity, most genome centers tend to specialize in two or three or four specific areas. A group at the University of Texas Health Science Center in San Antonio is building various maps of chromosome 3 and developing a chromosome 3 database, for instance. A cooperative led by the University of Iowa is creating a genetic linkage map for all the chromosomes. At Lawrence Berkeley Laboratory in Berkeley, California, researchers are developing robots to perform genome tasks and new computer systems to handle massive amounts of sequence information.

This chapter will look at two big centers. One is the Human Genome Center at Los Alamos National Laboratory, which is run by the Department of Energy. The other is the Genome Science and Technology Center at the Children's Hospital of Philadelphia, which is funded by the National Institutes of Health.

LOS ALAMOS NATIONAL LABORATORY

Although Los Alamos National Laboratory is perhaps best known for its weapons research, it is also home to one of DOE's genome centers. (The others are located at Lawrence Livermore National Laboratory in Livermore, California, and at Lawrence Berkeley Laboratory.)

Established in 1988, when DOE was first beginning its genome effort, the Los Alamos center drew upon its preexisting strengths in human genetics research. At that time, it was home to GenBank, a computerized database of all the DNA sequences that had been discovered and published in science journals. It had begun creating libraries of DNA fragments. It also had an active life sciences group.

Today, Los Alamos is one of the largest and most diverse genome centers in the country. It is dedicated to mapping chromosome 16. Its DNA libraries are sought by labs all over the country. Its computer informatics division—backed by the world's most sophisticated computer systems—is creating new software and hardware that is friendly to genome scientists. It even boasts a small ELSI (Ethical, Legal, and Social Implications) group. About thirty-five people work in the center.

Deputy director Larry L. Deaven, together with director Robert K. Moyzis, is responsible for promoting communication and collaboration at Los Alamos. How does he do it? How does he get biologists and physicists to talk to each other in a language each group can understand? How does he get mechanical engineers and laser scientists interested in the problem of DNA sequencing?

Deaven, whose background is in cell biology and genetics, makes it clear that he is not a multilingual interpreter. "In order to manage a multidisciplinary program like this you have to rely on advice," he says. "No one has a broad enough amount of expertise and experience to completely do it on his own."

— — — —

ROBOTICS

Take the robotics program, for example. The people involved with the DNA library project thought that a robot might come in very handy since preparing the libraries involves performing a number of steps over and over again. So they went and talked with the people in the robotics division, not quite knowing how or if the collaboration would occur. After all, robotics and life science had never collaborated before.

Luckily, someone emerged from the robotics group, says Deaven, who understood "the language and approach necessary a little bit better than the other people in robotics." That person took it upon himself to find out as much as he could about the biology project.

Although creating DNA libraries is painstaking work, the biology behind them is fairly straightforward. DNA fragments can be duplicated—"cloned"—to produce unlimited numbers of copies by inserting the fragments into a rapidly reproducing host cell. Once in the

— — — —

host cell—an *Escherichia coli* bacterium, for instance—the DNA fragment is reproduced right along with the host.

The library of an average-sized chromosome contains 3,125 fragments. (Each fragment comprises aproximately 40,000 nucleotide base pairs.) Because each fragment must be grown in its own host cell colony, preparing a copy of an entire library is very tedious. A portion of each of the 3,125 colonies must be transferred onto a new laboratory dish. Even with a special ninety-six-prong hand stamp to move ninety-six colonies at once, it's a very monotonous job.

Enter robot. The Los Alamos robotics team has designed a robot that can scan a bar code on a plate of bacterial colonies containing DNA fragments, dip its ninety-six-prong tool into the ninety-six wells on the plate, transfer the host bacteria onto a new plate, sterilize the tool, and begin again. This has proven invaluable to Los Alamos's mission of supplying other laboratories with DNA libraries. As of October 1995, the lab had sent out hundreds of DNA libraries copied by robots. Use of the robot is expected to speed production and eliminate error.

Other Los Alamos divisions have been similarly useful to the genome effort. People with backgrounds in chemistry, physics, and laser science have worked together to develop a radical new approach to DNA sequencing. An industry partner, Life Technologies, Inc., also contributes to this ambitious effort.

DNA SEQUENCING

Although this method remains untried, researchers at Los Alamos hope that one day they will be able to dye each nucleotide base a different color, clip the nucleotide bases off the DNA strand with a restriction enzyme, and float the bases in water past a laser detector. The detector

At Los Alamos National Laboratory, a robot dips a ninety-six-prong tool into a bacterial colony containing DNA fragments from chromosome 16.

would read the different colors of the bases and send that information to a computer.

If this ambitious effort proves successful, says Deaven, the lab will be able to sequence 1,000 base pairs per second. Obviously such productivity would be invaluable to a project that plans on identifying the sequence of the 3 billion base pairs in the human genome.

Recording the DNA sequence of the human genome would fill 1,000 phone books that are each 1,000 pages long. New computer systems are needed to handle the volume of the information on DNA sequences and gene locations that is pouring forth from the project. Computer scientists are a crucial component of the genome effort.

Computer programmers are useful in another way too. Programmers don't think like biologists when they look at long DNA sequences. They think like, well, programmers and they tend to view the biological code as similar to a computer programming language like FORTRAN or BASIC. This is a potentially useful way to approach the non-protein-coding regions for which we haven't yet broken the code.

Even the protein-coding regions for which we do have the code are not simple to read. For one thing, it's difficult to decide where to start reading the nucleotide triplets. For example, how would you read this string of bases: CGCCTCGGCCTCTG? Is it:

CGC   CTC   GGC   CTC   TG _ ?
arginin   leucine   glycine   leucine

Or: **_ _ C   GCC   TCG   GCC   TCT   G _ _ ?**
alanine   serine   alanine   serine

Or: **_ CG   CCT   CGG   CCT   CTG ?**
proline   arginine   proline   leucine

The resulting protein will be very different depending on how you read the sequence. Computers can be programmed to try different reading strategies and to decide which one is most likely to be correct. Certain triplets, known as "stop" triplets, mark the end of protein-coding regions. Computers will look for stretches that aren't interrupted by stop triplets. This is one way computers can be used to distinguish between protein-coding and non-protein-coding regions.

CHILDREN'S HOSPITAL OF PHILADELPHIA

On the other side of the country, Beverly S. Emanuel is in charge of a big center at the Children's Hospital of Philadelphia that is devoted to mapping and sequencing chromosome 22. This center, established by NIH in 1991, grew out of Emanuel's long-standing work with chromosome 22, especially the collaborations and resources her group had developed prior to the Human Genome Project.

Every chromosome has its own story, its own set of intriguing genes, and chromosome 22 is no exception. In 1974, as a young molecular biologist, Emanuel came to the hospital and developed its chromosome testing lab. She soon noticed that a disproportionately large number of abnormalities were associated with this small chromosome. For two decades now, her work has focused on those abnormalities, many of which occur within a particular region of the chromosome.

TRANSLOCATION

Some of the abnormalities are the result of two chromosomes swapping pieces. This kind of genetic mistake is known as a translocation. A translocation between chromosome 22 and chromosome 9 causes a cancer known as chronic myelogenous leukemia. (This translocation, dis-

Beverly S. Emanuel pioneered research on chromosome 22.
Mutations of chromosome 22 can lead to cancer or heart defects.
She is the director of the Genome Science and Technology Center
at the Children's Hospital of Philadelphia.

covered in 1960 in Philadelphia, is known as the Philadelphia chromosome.) A translocation between chromosome 22 and chromosome 11 causes a bone cancer known as Ewing's sarcoma. All told, at least eight types of cancer have been linked to defects on chromosome 22. Evidently, if certain genes find their way to the wrong chromosome, they trigger cancer.

Like Joan Overhauser at Thomas Jefferson University, Emanuel is also investigating several syndromes involving chromosomal deletions. DiGeorge syndrome, marked by developmental defects and mental retardation, is caused by a deletion on chromosome 22. Emanuel has also discovered that a large percentage of children born with heart defects have chromosome 22 deletions. Cat Eye syndrome, so named because the people with this syndrome have eyes like cats, is the result of a duplication of chromosome 22.

Healing sick children is a big part of what motivates Emanuel. Her office overlooks the entrance to the Wood Pediatric Center of the hospital. Downstairs are two large waiting areas filled with entertainment for children: large televisions, colorful plastic slides, and miniature washer/dryer sets. The connection between laboratory research and patient care is very close here at the Children's Hospital.

From the beginning, Emanuel's research has been guided by one question: Why is this region of chromosome 22 so susceptible to rearrangement? As the molecular tools became available, Emanuel and her colleagues began mapping chromosome 22. Then along came the Human Genome Project.

When the project was announced, Emanuel's lab already had a library of DNA clones and other resources for mapping chromosome 22. Her lab had a core team of diverse and committed investigators. Emanuel's team also actively collaborated with several other labs. "We knew from the inception of the Human Genome Project that we had something useful to offer," she recalled. Yet with the $10 million NIH awarded her to establish a genome center, she was able to strengthen her group even further.

Emanuel's center has two primary goals. First, it is devoted to mapping chromosome 22. With its large talent base, the center has the luxury of developing physical and genetic linkage maps simultaneously. As you recall, the two maps measure distance between genes in a different way. The *order* of genes is the same, but the distances look different. This is because recombination, the event measured by the genetic map, is not equally likely at all places on the chromosome. As a result, the genetic linkage map is stretched in some places and squished small in others. By periodically reconciling the two types of maps, an accurate final map can be created.

The center's second focus is developing computer systems for genome scientists. A new, large group of computer scientists and computational biologists hired specifically for the genome center is working to construct data management tools and DNA analysis programs.

GROUP EFFORT

The Genome Science and Technology Center at the Children's Hospital is really a collection of the talents of thirty people. Many members of the center are located at the hospital, of course. But some of the key researchers are at the University of Pennsylvania Medical Center. This is somewhat different from the genome group in Los Alamos, where everybody is an employee of a national laboratory and working under one administrative roof.

The genome center designation has also led Emanuel into some new collaborations—and interesting introductions. Soon after her team was declared a center for chromosome 22, Emanuel established an electronic "bulletin board" for chromosome 22 that could be read by anyone with a personal computer and modem. A message from scientist Bruce Roe at the University of Oklahoma appeared. He announced that he had received NIH funding to sequence biologically interesting areas on chromosome 22. "Does anybody have any?" was the end of his message, as Emanuel recalled.

Well, she did. But before sending her clones, she checked to see whether this fellow was on the level. He was indeed, and Roe is now one of Emanuel's collaborators, doing large-scale DNA sequencing.

"I cannot imagine this would have happened without the infrastructure and heightened awareness that the Human Genome Project brought to mapping chromosomes," Emanuel said. Collaborative efforts like these, crucial to finishing the genome project, are especially easy to see in the big centers. It will be interesting to see whether large collaborations like these will have a place in biology once the genome project is finished.

Los Alamos's Larry L. Deaven said that he can already see a difference in the biologists at his center: "The new young students tend to put more emphasis on math and physics than before. It's already a trend. Students are emerging from Human Genome Project centers combining different sets of skills. A new kind of scientist is emerging."

MICE AND FLIES UNDER INVESTIGATION

Human Genome Project might not be the best name for this endeavor. Why? Because human DNA is not the only DNA under investigation. Five other species, including the mouse, are also part of the project. This underscores an important point: The building blocks of life—nucleotide pairs, DNA, amino acids, and proteins—are very much the same no matter where they appear. The same building blocks are used to construct humans, whales, and mice. It's just that the amounts and combinations are somewhat different (although not as different as you might think!).

MODEL ORGANISMS

The goals for the model organisms are identical to those for the human genome: creating physical and genetic maps, and sequencing DNA. The laboratory mouse, the bacterium *E. coli*, the fruit fly *Drosophila melanogaster*, the roundworm *Caenorhabditis elegans*, and the yeast *Saccharo-myces cerevisiae* are the species under study. Genome research with these model organisms are progressing very well.

Why bother mapping and sequencing the genomes of other species? For three reasons. First, the genomes of all but the mouse are easier to work with than the human genome. Second, identifying genes and their function in other species will help pave the way for interpretations in the human genome. Third, the study of model organisms leads to a better understanding of human disease.

SIMPLE ORGANISMS

Simple organisms make good models with which to develop new technologies. Their DNA is shorter and easier than that of humans. Single-celled species like *E. coli* have only protein-coding DNA (exons), for instance. They do not have the long, confusing, non-protein-coding regions (introns) that make up so much of human DNA.

This is especially helpful for scientists when it comes to sequencing. In fact, scientists at the University of Wisconsin hope to finish sequencing all 4.7 million base pairs of *E. coli* by 1998. Ideally, any technology developed with simple organisms can be transferred to the study of more complicated organisms.

UNDERSTANDING THE GENOME

Human genetics has long benefited from the study of other species. Gregor Johann Mendel's famous experiments with heredity were conducted with pea plants in the mid-nineteenth century, after all. Fruit flies, which reproduce very quickly, are excellent genetic models. Laboratory experiments with fruit flies, conducted since the early 1900s, have vastly increased our knowledge of

The fruit fly is one of five nonhuman species that the Human Genome Project is studying. Many kinds of genetic mutations in fruit flies, including variations in eye color and wing length, have been observed.

the causes of mutation. Scientists Joshua Lederberg and Edward Lawrie Tatum discovered the phenomena of genetic crossing over in 1947 by studying *E. coli*.

Such research points to underlying biochemical mechanisms that are universal to all species. Biologists who have charted the development of the roundworm have discovered the origins of each of the 958 cells in the adult animal. This kind of investigation into the developmental biology of a simple animal helps researchers ask intelligent questions regarding the development of a more complex animal, such as a human.

How are the biochemical reactions that are present in a developing roundworm similar to those in a developing human? How are they different? What genes are responsible for the different stages of development? Answers to such questions may offer new insights into both fetal development and the aging process.

Identifying a gene is only the first step in understanding its protein product. Even knowing a gene's sequence—and, thus, its amino acid order—does not tell scientists much about the protein it codes for. That's because the three-dimensional shape of a protein is as important as its biochemical structure. Furthermore, the protein product of a gene may be present only in certain organs or at a certain stage of fetal development. Indeed, isolating a gene is often much simpler than figuring out how it functions within an organism.

This is where the study of model organisms is valuable. Let's say you've found a gene in a region of chromosome 18 or 22 that is deleted in people with chromosomal deletion syndromes. You know the DNA sequence of your gene. You take that sequence and enter it into a computer database containing the sequences of known human proteins. No match.

Not yet discouraged, you enter your sequence into a database of sequences of genes from yeast or fruit flies or mice. You're in luck! It matches a yeast gene that codes for a protein involved in cell structure. Or maybe it matches a mouse gene that causes uncontrolled cell growth—cancer—under certain conditions.

In this fashion, hundreds of genes have been identified in humans. As a rule, the likelihood of making a match with a D. *melanogaster* or mouse database is greater than the chance of making a match with a human database, simply because those model organisms have been more thoroughly studied.

SHARED DNA

It may be difficult to believe that humans and mice share any common DNA. It may be even harder to believe that humans and yeast cells do. Yet measuring the amount of DNA shared between two species is a way of measuring how closely related two species are in terms of evolution. Species that are somewhat similar share many genes. Humans and mice share DNA that is unique to vertebrate animals. Humans and fruit flies share DNA that is essential to all multicellular organisms. Humans and yeast share DNA that is essential to the life of all cells with a nucleus. As you can see, the degree of relatedness between humans and other species becomes more basic the farther down the evolutionary tree you go.

It may seem that a human and a mouse are worlds apart. And they are, if you're just thinking about different animals. But if you begin thinking across the entire spectrum of living organisms—algae, fir trees, bacteria, snails—then suddenly mice and humans have more in common than you might think.

The mouse is a particularly important model organism because it is so closely related to humans. About 90 percent of the mouse genome is identical to that of the human genome. Many mouse genes, presumably those genes that govern mammalian development, are identical to human genes. For this reason, the Human Genome Project has placed a special emphasis on sequencing the mouse genome side by side with the human genome.

This approach might be especially useful for distinguishing between exons and introns. It will also give us a better understanding of evolution. Those regions of DNA that are common to humans and mice have managed to survive for millions of years. They are referred to as "conserved" regions. As you might expect, mapping and sequencing the mouse genome will be nearly as challenging as mapping and sequencing the human genome.

HUMAN DISEASE

The treatment of human disease has long depended upon the use of laboratory animals. While scientists have developed many medicines for humans by testing potential cures on animals first, animal rights activists generally oppose the use of animals in research.

Research of genetic diseases—especially those illnesses that seem to have both an environmental and genetic component, such as diabetes and cancer—will be greatly aided by the study of genetic disease in lab animals. Lab mice have already been genetically altered to study sickle cell anemia, dwarfism, muscular dystrophy, and other diseases or conditions.

Much of the work on model organisms is being done at NIH Genome Science and Technology Centers. Five of the NIH's eighteen centers are devoted to the study of model organisms. It's a logical way to organize a center, after all. The Department of Energy is not involved with model organism study.

For the most part, the centers were created where a team of university researchers already had an established genome program. The genome of the fruit fly is under investigation at the University of California, Berkeley, for example, while the roundworm is being examined at Washington University in St. Louis. A team at the Whitehead Institute for Biomedical Research in Cambridge, Massachusetts, is devoted to studying the mouse. Not all genome project work on other species takes place at large centers, however. Many small teams at institutions around the country are also studying these organisms.

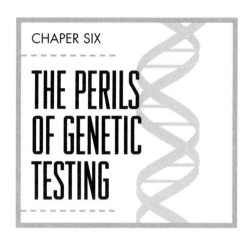

THE PERILS OF GENETIC TESTING

Twenty-six blocks west of Joan Overhauser's laboratory, a short bus ride past Philadelphia's City Hall, over the Schuylkill River and to the edge of the campus of the University of Pennsylvania, sits David A. Asch's office.

Asch, too, is working on the Human Genome Project. Yet unlike Overhauser, he has no shiny lab, no busy team of graduate students and technicians. He is not studying chromosomes. Instead, he is examining how the genome project may affect our lives.

His office is on the third floor of Ralston House, a graceful brick building that was once a private "Home for Aged Women." The building's double staircase and small piano harken back to an era when things like genetic screening and prenatal testing were unheard of.

Asch is examining how society might choose to use the information that the Human Genome Project is churning forth. In one of his projects, he is examining the implications of the discovery of the cystic fibrosis

David A. Asch, a physician and professor at the University of Pennsylvania, examines how the discoveries of the Human Genome Project are being used. Many people wonder whether the disadvantages of genetic testing might outweigh the benefits. Asch's study of the screening test for cystic fibrosis is designed to help physicians and parents use the test wisely.

gene. Discovering the gene's location made it possible to develop a test for the mutation. That test now exists. Used in connection with *amniocentesis*, a procedure in which fetal cells are drawn from the fluid in the sac surrouonding the fetus, expectant parents can learn whether their fetus has inherited two copies of the disease gene. Used on adults, the test can show whether a healthy person carries the defective gene. An additional concern is that 1 in 200 amniocenteses causes miscarriage.

There is just one problem. The screening test is not completely accurate. Sometimes it misses people who are carriers. The current version of the test can identify only about 85 percent of Caucasian carriers. In other words, 15 percent of those who test negative really are not. In actuality, they are carriers of the disease gene.

The reason for this is that more than one genetic defect can lead to cystic fibrosis. Seventy percent of people with cystic fibrosis have the phenylalanine 508 deletion that Tsui's and Collins's teams discovered. But disease in the other 30 percent is caused by any one of 300 different mutations in the gene. At this point, no screening test, no matter how sophisticated, can test for all the defects. No test can guarantee, with 100 percent accuracy, that a person is *not* a cystic fibrosis carrier.

Asch is exploring the different ways that parents and health professionals might use this imperfect screening test. He is considering all the various choices facing parents and health professionals who might wish to use the screening test.

Think about it. If you were a doctor, how would you use the test? Imagine that a couple expecting a baby and concerned about cystic fibrosis came to you. If both parents test postive as carriers, their baby has a one in four chance of inheriting two copies of the disease gene. (Remember that a person must have two copies of the disease gene to develop cystic fibrosis.) But if one or both parents tests negative, you can't completely trust the results. The screening test might have missed their particular mutation, and they could still have a baby with cystic fibrosis.

What would you do? Would you recommend the test? Would you test both parents or only one?

These are exactly the kinds of questions that the people behind the Human Genome Project wish to have asked. They know that genetics research can be greatly misused. You only need to look as far as the Nazis' desire for "genetic purity," which they used to justify the murder of 6 million Jews during World War II. Gypsies, homosexuals, and the mentally ill were also killed. Even in the United States, a sickle cell anemia screening program for African-Americans was badly bungled in the 1970s. Many people who were healthy carriers met with discrimination from employers and life insurance companies.

ELSI BRANCH

From the beginning, the Human Genome Project has taken steps to prevent the misuse of genetic information. The project spends 5 percent of its annual budget on the study of the ethical, legal, and social implications (ELSI) of its research. It is the first time that a large research project has devoted money to examining its own effects. Asch is one of the first researchers to receive funds from the ELSI branch of the genome project. A look at his work will provide a glimpse into the larger ELSI effort.

Asch's research studies several cystic fibrosis screening strategies. Each has different advantages and drawbacks. One approach, for instance, results in the fewest number of cystic fibrosis births, but it is also costly and results in many abortions and miscarriages of healthy babies. (It's impossible to discuss prenatal testing without acknowledging that it can lead to abortion. This is one reason why it's such a loaded ethical issue.) Other strategies have other outcomes.

Research like this requires a complicated mix of talents: medical knowledge, people skills, and an understanding of how health care is delivered. Asch, a physician with an undergraduate degree in philosophy and a master's degree in health care administration, is uniquely qualified for the challenge.

Together with three colleagues who had similar expertise, Asch reported on several models for cystic fibrosis screening in an article published in the January 1993 issue of the *American Journal of Obstetrics and Gynecology.*

"The purpose of these models is to get an advance indication of how various ways of using these tests will affect society," Asch says. He is young—in his early thirties—and speaks quickly and enthusiastically. "They're helping to predict the future. You try to follow a strategy to its logical or illogical conclusions and see what happens. You see how many kids are born with cystic fibrosis, how many abortions occur, how many deliveries of unaffected kids, how many miscarriages result from amniocentesis, and how much money is spent." Money spent on genetic testing includes not only the cost of the DNA tests, but also the cost of taking time off work, hiring baby-sitters, paying for parking, and so on.

The models are simplified representations of the real world. They can be used by expectant couples to view the consequences of a particular path of decisions.

Here's an example of one decision path. Pretend, for a moment, that it's your decision path. You're a woman who is pregnant with your first child. You decide to get tested for the cystic fibrosis mutation. Why not? What could be the harm? Besides, the odds are only 1 in 625 that both you and your husband are carriers.

At your next prenatal visit, the nurse takes a blood sample and charges you several hundred dollars for a DNA test. The results surprise you. You're a cystic fibrosis carrier. You had been told that 1 in 25 people of European ancestry is a carrier, but you never thought it would be you. Now the genetic counselor recommends that your husband get tested. Bad news again—he is a carrier too.

Now you're frantic. The odds are 1 in 4 that the fetus inherited two defective copies of the gene, one from you and one from your husband. Your doctor recommends an amniocentesis to examine the baby's DNA. You agree, even though it's expensive and the procedure may cause a miscarriage. You take a day off from work to get the amniocentesis done.

When the results come back, you learn that the fetus inherited two flawed genes. Now you're really upset. What should you do? Your best friend urges you to abort the fetus because it isn't right to bring a sick child into the world. You had always thought that you would abort a fetus with cystic fibrosis, but now you're not so sure. Your sister tells you that one of her friends has cystic fibrosis and lives a fairly normal life. Your husband, also upset, says the decision is yours.

You ultimately decide not to terminate the pregnancy. About 5 months later, you deliver a baby boy with cystic fibrosis. The outcome of this pregnancy is exactly the same as if you had never been tested, never had the amniocentesis, never taken the time off work, never gone through all the mental anguish.

In his models, Asch traces the possible outcomes of various choices in a step-by-step manner. His models are useful because consequences that are not especially obvious suddenly become glaringly clear. Even though all these tests sound good in theory, the possibility of inaccurate results makes them less advantageous than they may initially seem.

EUGENICS

In the future, it may be possible to cure cystic fibrosis. It may even be possible, one distant day, to treat cystic fibrosis in the womb. As of 1995, gene therapy for the

A genetic counselor and a couple look at a karyotype together. Genetic counselors help couples make decisions about prenatal testing. As the number of genetic tests grows, so will the need for genetic counselors.

disease looked promising. But it was still only in the most preliminary stages. In the meantime, couples who learn that their unborn child has cystic fibrosis have only two choices: carry the pregnancy to term or abort.

Prenatal diagnosis, then, might seem to suggest a willingness to choose abortion. It might seem, too, that people who are opposed to abortion would never seek prenatal testing for cystic fibrosis or other diseases. But for the vast majority of people, prenatal testing ends with happy results. It provides enormous reassurance that a healthy baby will be born. Even people opposed to abortion might want prenatal tests done for the peace of mind they're likely to receive.

Yet, as might be expected, antiabortion activists are still highly critical of this aspect of the genome project. One activist has even called the project a "search and destroy mission."[1]

Prenatal testing also sends a clear message to people with cystic fibrosis. Asch quotes several young people with cystic fibrosis who have told him that they know why prenatal testing is done: "It's so that people like me aren't born." It's a painful thing to hear, and Asch's voice rises as he relays this comment. Suddenly Asch, wearing a casual sweater and surrounded by pictures of his family, seems more like a dad than a doctor.

"It raises questions of eugenics," Asch adds. "Are cystic fibrosis kids not worth having?" (Eugenics is the attempt to improve a population by encouraging the "fit" to have children and the "unfit" not to. The word eugenics was coined in 1883 by Francis Galton, a cousin of Charles Darwin.)

Indeed, the Cystic Fibrosis Foundation, an organization devoted to improving the lives of cystic fibrosis patients through research, opposes genetic testing. Rather than encouraging efforts that might prevent cystic fibrosis babies from being born, it is dedicated to improving the lives of children and young adults who have the disease.

"Cystic fibrosis is a situation where molecular biology has a tremendous amount to offer," Asch says. "Still, it's fraught with all sorts of problems. Most of the other genetic tests coming down the pike are going to be significantly worse than this in terms of accuracy. The diagnosis is going to be so much more complicated either

76

because there's many more mutations, many genes involved, or maybe a combination of genetic and environmental factors at work."

So what's the answer? Should genetic testing be made available? Asch and others working within the ELSI branch of the Human Genome Project can provide information that helps people think about this question. But ultimately ethical discussions of this sort demand thoughtful consideration and debate in classrooms and around dinner tables across the country. Is a genetic test with some degree of inaccuracy acceptable? Should every pregnancy be screened for genetic defects? Should couples be tested for carrier status before they start a family?

Questions like these are too important to be left to the experts.

CHAPER SEVEN

THE BIG PICTURE

The purpose of the Human Genome Project is to develop innovative tools that will help scientists find genes and explore the function of DNA. These tools include maps of the genome, technology for DNA sequencing, and improved computer databases to hold the information. The genome project officially began on October 1, 1990, and is scheduled to run for 15 years. Its deadline for completion is September 30, 2005.

And *whose* genome is this, you ask? Not that of one particular person, actually. The DNA studied will be taken from many different people. All the samples will contribute to a representative human genome. Think of medical students studying a human heart. They don't ask *whose* heart it is. The students study this model heart, fully aware that some people have larger hearts, some have smaller hearts, and some people have other variations. All hearts share the same important features, however. Likewise, the human genome is 99.5

percent the same in all of us. In fact, to find the 0.5 percent difference, we'll first need to identify the similarities.

The cost of the Human Genome Project's total 15-year budget has been estimated at $3 billion. Mapping and sequencing the human genome by 2005 was the original goal of the genome project. The first 5 years of the project have seen considerable progress in genetic mapping. Physical mapping is also going well, and maps of the genome will probably be completed before 2000. Sequencing the human genome has been much slower, however, and project leaders say that new technology must be develped to finish the job by 2005.

Meeting the project's goals will also require advances in a number of areas. As a result, the Human Genome Project is supporting a variety of projects related to its main goal. The work of analyzing the human genome, for instance, is made much easier when the genomes of other species are known. The gathering of nucleotide sequences and other information—an enormous amount of data—requires the development of new computer tools and information systems. The genetic knowledge that is coming will raise ethical, legal, and social issues. Young people just beginning their educations and careers need to be encouraged to work on the genome project.

SETTING GOALS

To ensure that programs related to the Human Genome Project receive the support they need, the project organizers have set goals in seven important areas:

- The Human Genome: Map and sequence the human genome with an emphasis on identifying genes.
- Model organisms: Map and sequence the genome of five model organisms (the laboratory

mouse, *E. coli* bacteria, the roundworm *C. elegans*, the yeast *S. cerevisiae*, and the fruit fly *D. melanogaster).*

- Ethical, Legal, and Social Implications: Identify issues in an effort to anticipate and plan for problems before they arise.
- Information Handling and Analysis: Develop database systems for data collection and analysis that will allow researchers from all over the world to use the genome project findings.
- Technology Development: Support improvements in the methods used for genome study, especially DNA sequencing.
- Technology Transfer: Support the transfer of project technology into industry and other areas where it may be of use.
- Training: Encourage students and scientists to become trained in the skills needed for genome research. The project relies heavily on people trained in more than one discipline.

These are the general goals that will see the project through its 15 years of existence. Five-year goals in each category have also been established. The first set of 5-year goals was published in 1990. Unexpected (yet very welcome) advances prompted the project leaders to publish another 5-year plan in October 1993. These goals will need to be updated in September 1998.

If you think in terms of these goals, the big picture of the genome project begins to form. You can start to see where the work of Overhauser, Asch, and the others fits in. Joan Overhauser, with her study of chromosome 18, is contributing to the project's effort to map the genome. David Asch, by studying cystic fibrosis screening, is examining the ethical, legal, and social implications of the new genetic discoveries. Research training, another aim of the genome project, occurs whenever graduate students work in the laboratories of scientists funded by the project. Graduate students often work in the labs of Overhauser and Emanuel, for instance.

FOCUSING ON THE BIG PICTURE

So, is anyone responsible for this big picture? Yes, of course. The director of the National Center for Human Genome Research, at the National Institutes of Health, is Francis Collins. The person responsible for the Department of Energy's genome effort is Aristides Patrinos. Patrinos is the associate director of the Health Effects and Life Sciences Research Division in the Office of Health and Environmental Research.

Collins, you may recall, was one of the codiscoverers of the cystic fibrosis gene in 1989. After that victory, his team went on to discover the gene for Huntington's disease and neurofibromatosis. In 1991, his group at the University of Michigan was designated an NIH genome center, like that of Beverly Emanuel. When James Watson resigned as director of the NIH's genome effort in 1992, Collins was offered the position.

On an unseasonably hot spring day in 1994, in his office on the NIH campus in Bethesda, Maryland, Collins speaks about the challenges of his job. "One of the appeals of this position is that there is such an impressive big picture," he says. "In a given day, I might go from a very intense discussion about the new miniaturized technology that will potentially allow us to do DNA sequencing on a credit card, to a debate about the ethical consequences of presymptomatic testing for colon cancer, and then to my lab here where somebody will show me their gels and talk to me about the experiment they did that morning and why it worked or didn't work."

When *Time* profiled Collins, it described him as a "workaholic who logs 100-hour weeks."[2] Clearly the job requires such dedication. Yet in other ways, Collins is hardly the stereotypical scientist. The photograph that ran with the *Time* article shows him in black leather and boots sitting on a Honda Nighthawk 750 motorcycle. (He keeps track of each gene he has discovered with stickers stuck to one of his helmets.) He is also a devout Christian who has twice traveled to Nigeria to work with missionary doctors.

One of Collins's responsibilities as director is to keep the project running smoothly and on schedule. "What's missing in our ability to get this project done by 2005?" he asks, demonstrating the kind of question that fuels his agenda. "Some of the pieces are there, but what about the ones that aren't there? And how are we going to get them there? And how are we going to recruit into this project people who wouldn't have thought of themselves as geneticists—or even biologists!—to help us deal with the technical challenges of getting the job done?"

The main duty of Collins and Patrinos is to stay focused on the primary goals of the project. At the same time, the two leaders are very much aware that the Human Genome Project is not only an end in itself, but also a means to an end. Think of it. If the project is successful, we will know the complete nucleotide sequence of the human genome in ten years. We will know the location of all the genes. An astounding achievement, worthy of celebration. Yet much will still be unknown.

To know the gene means knowing the protein that it codes for, as we saw. But that isn't the same as understanding the complex biochemical steps that make the body function. By the turn of the century, Overhauser might very well know which proteins are implicated in chromosomal 18 deletion syndrome. It's unlikely, though, that she will understand precisely how the absence of those proteins causes the symptoms she sees.

The completion of the Human Genome Project will provide a sourcebook of information for scientists. This sourcebook will be available on computer. It will also be available in the form of laboratory dishes that contain DNA segments. These identified, labeled sequences of DNA will be available to other labs for research.

The completion of the genome map will be an enormous boon to biological research. It will also allow us to investigate questions that were never possible to investigate before: How do we grow from fertilized cell to adult? What turns a gene "on" and "off" so that it knows whether to create a brain cell or a blood cell? What is memory? Why do we grow old? We will also be able to start moving toward a better understanding of how environmental factors—the food we eat, the air we breathe, the work we do—affect our DNA and our odds of developing cancer and other diseases.

"The Human Genome Project to my mind is really a historic undertaking," Collins says. He notes that basic scientific research will benefit enormously. "But the real reason the public is paying for it is that they have been led to believe—by myself and others—that this is going to benefit human health."

THE ROLE OF PHARMACEUTICAL COMPANIES

The benefits to human health will come one step at a time, hopefully with a cure for genetic disease as the final step. When a disease gene is discovered, pharmaceutical companies begin to get involved. The first response of these companies is usually to develop a screening test for the disease gene. Their next response is often the research and development of new drugs that compensate for missing proteins.

Developing and marketing screening tests and drugs is the proper role for commercial companies. Although many pharmaceutical companies exist, few have the money and talent for genome mapping. Research scientists and their institutions, on the other hand, lack experience and interest in drug development. Modern medicine depends on both groups. Yet it's often difficult to strike the right balance between commercial enterprise and research science.

The reason is that companies need to carefully protect their investments. They usually do this by patenting their discoveries or keeping them secret from their competitors until they are ready to reveal them. Scientists, on the other hand, are accustomed to sharing information with others by publishing the results of their work in science journals. This free exchange of information has traditionally been seen as crucial to the advancement of science.

Early in its history, the Human Genome Project was at the center of a disagreement between commercial and research interests. In 1991, under director Bernadine Healy, NIH filed for patents on 6,122 gene fragments produced by the genome project. The agency wished to benefit financially if any of those fragments proved useful to the development of drugs or screening tests in the future.

But many scientists saw the patent application as ridiculous. Patenting a DNA segment was hardly the same as patenting a new drug. Furthermore, how could NIH claim exclusive rights to the genetic inheritance all humans share? Ultimately, the U.S. Patent and Trademark Office sided with these views. In 1992 and 1993, the office rejected all of NIH's patent applications.

The information gained by deciphering the code of the human genome is potentially relevant to everyone. It stands to reason, then, that the United States would not be the only country involved in such a project.

While the United States has put more money and effort into its genome project than other countries, scientists around the world are interested in mapping and sequencing the genome. Progress in genome research has been made in the United Kingdom, France, Italy, and elsewhere in the European Community, as well as in

Victor McKusick, a medical geneticist at John Hopkins University, was the first president of the Human Genome Organization (HUGO). The international group is dedicated to keeping genome information free and available to scientists around the world.

Japan, Russia, Canada, and a network of Latin American countries. Germany, although very strong scientifically, was slow to mount a genome project because of its Nazi past.

Two private groups are helping to support international genome project efforts. The Howard Hughes Medical Institute helps fund a database at the Center for the Study of Human Polymorphisms in Paris, for instance. This database contains genetic linkage information on a number of large families. Founded by aviator (and billionaire) Howard Hughes, this private organization also supports biomedical research in the United States.

The Human Genome Organization (HUGO), incorporated in Geneva, Switzerland, is a kind of United Nations for the human genome, with 220 member scientists from twenty-three countries.[3] Organized in 1988, its founding president was Victor McKusick, an American geneticist. HUGO works to keep genome information freely accessible to scientists around the world. The group helps fund the international chromosome workshops, like the ones in which Joan Overhauser and Beverly Emanuel participate.

The Human Genome Project was never a sure thing. It did not happen overnight. That it even exists at all is testimony to a number of dedicated and visionary people. Since its birth in the mid-1980s, the project has changed shape and direction many times. A look at those early years—marked by criticism and compromise—will cast some light on its current status. It will help explain why two government agencies oversee it, and why its research takes place in both small labs and large centers.

CHAPTER EIGHT

OF ATOM BOMBS AND CANCER

In August 1944, the United States dropped two atomic bombs on Japan, one on Hiroshima and one on Nagasaki. Shortly afterward, the Japanese surrendered and World War II ended. The bombs killed more than 100,000 people instantly. Many others suffered radiation poisoning. No one knew what effect low levels of radiation had on the DNA of those who survived the blast.

The A-bomb, built at Los Alamos National Laboratory under the supervision of the Department of Energy, defined our century. Its existence also led, in an admittedly indirect way, to the Human Genome Project.

Thirty years later, the United States declared war on a more intimate enemy: cancer. In 1971, President Richard M. Nixon signed the National Cancer Act. The act provided funding for scientists struggling to pinpoint the root causes of cancer. What made certain cells reproduce wildly and form tumors? What genetic errors and environ mental exposures set those cells on their lethal course? A better understanding of DNA would surely help answer such questions.

In 1973, the biotechnology revolution began when Stanley Cohen of Stanford University and Herbert Boyer of the University of California San Francisco, used enzymes to cut and slice DNA and to move DNA from one organism to another. In 1980, David Botstein and Ronald Davis developed a new strategy for hunting unknown genes by using DNA markers and genetic linkages.

These advances provided scientists with everything they needed to explore the human genome. Just as important, the Department of Energy had the political savvy to see that it was the right moment in history to propose such an ambitious undertaking.

Such are the historical roots of the Human Genome Project. At no time was it ever the vision, the sole property, of a single individual. Rather, it was born and nurtured at several scientific gatherings where investigators came together and swapped seemingly outrageous ideas.

THE ALTA SUMMIT

The idea of a genome project seems to have emerged simultaneously from several different meetings and conferences held in the early 1980s. One was a meeting of nineteen top molecular biologists sponsored by DOE and the International Commission for Protection Against Environmental Mutagens and Carcinogens. It was held in Alta, Utah, in December 1984. One of the DOE's areas of responsibility, as designated by Congress, was to monitor the effects on human health of various forms of energy, including nuclear energy. DOE organized the Alta meeting to talk about the effect of radiation and chemicals on human DNA.

The meeting organizers posed a question to the scientists they had invited to Alta: Was it possible, using the new DNA tools, to measure the damage done to human DNA by very low levels of radiation or chemicals? Although the DOE specifically referred to the children of Hiroshima and Nagasaki survivors, it was also addressing the concerns of Vietnam veterans exposed to Agent Orange, and of residents of Nevada and other western states who were exposed to radiation during atomic bomb tests in the 1950s.

Were the new DNA tools sensitive enough to ferret out thirty nucleotide base mutations in a single human genome? DOE asked. (That's one nucleotide base mutation in every 100 million.) No, replied the scientists. But they began to imagine the technology and information that would be needed to make such an effort possible.

A draft version of the meeting's report, *Technologies for Detecting Heritable Mutations in Human Beings*, reached Charles DeLisi, the director of the Office of Health and Environmental Research at the DOE, in 1985. While reading the report, DeLisi could see the advantage of developing technology that would quickly and cheaply sequence the human genome. If the human genome were known, even tiny variations caused by mutagens could be measured.

The Department of Energy already had a large group of biologists working within the national laboratories. It housed the world's most sophisticated computer systems. It was home to Genbank, a DNA sequence database, and to the National Gene Library, which made fragments of DNA available to scientists. Furthermore, the agency was accustomed to directing large, multidisciplinary science efforts. In this light, the project to sequence the human genome seemed a natural extension of the DOE's mission. DeLisi, with the clout of DOE behind him, quickly emerged as one of the major proponents of the Human Genome Project.

Some observers have questioned the timing of the

*Robert L. Sinsheimer, as chancellor of the University of
California, Santa Cruz, envisioned creating a center
dedicated to the study of the human genome on his campus.
In 1985, he organized one of the first meetings on
mapping the human genome.*

DOE's interest in applied biology projects.[4] The Cold War with the Soviet Union, sustained by stockpiles of nuclear weapons, was coming to an end. Perhaps the DOE, realizing that its weapons programs might soon be falling out of favor, needed to give itself a new reason to exist.

At about the same time, Robert L. Sinsheimer, chancellor at the University of California at Santa Cruz, also hit upon the idea of a large-scale genome project. Both a biologist and a university administrator, he had observed astronomers and physicists unite in support of large projects that they believed necessary to their work.

THE SANTA CRUZ WORKSHOP

In May 1985, just months after the Alta Summit, Sinsheimer organized a small meeting of scientists to discuss establishing a large-scale genome project on his campus at Santa Cruz. Three of the researchers who attended the Alta meeting were also present at Sinsheimer's Santa Cruz Workshop. Out of that meeting came a short appendix of goals for mapping and sequencing.

The next shot of enthusiasm came ten months later. On March 7, 1986, *Science* published an essay by Renato Dulbecco, Nobel laureate in medicine, titled "A Turning Point in Cancer Research: Sequencing the Human Genome." In this influential article, Dulbecco described how cancer researchers had begun to identify malfunctioning genes that could eventually lead to cancer. Further knowledge of human DNA would give cancer research "a major boost," Dulbecco observed. He compared mapping the human genome to the conquest of space and strongly urged his peers to support such an undertaking.

That year marked two critical meetings that independently led to the Human Genome Project. In March 1986, shortly after Dulbecco's article ran, DOE sponsored an international workshop in Santa Fe, New Mexico, home of the Los Alamos National Laboratory. The meeting was a success, and DOE began planning its genome project in earnest.

Two months later, Cold Spring Harbor Laboratory, a private research institution directed by James Watson, held a symposium on the molecular biology of humans. The symposium became an unexpected opportunity for molecular biologists to debate the merits of the genome project as proposed by DOE. Watson, although a powerful supporter of the project, objected to DOE's involvement. He believed NIH, the preeminent sponsor of biomedical research, should lead the effort. Other biologists simply disliked the idea of a large, targeted project with a definite endpoint. They feared it would draw money away from their small research teams.

Many of the concerns that surrounded the genome project's early days were aired at this meeting. Indeed, much of the project's early life was shaped by three debates: big teams versus small teams, sequencing versus mapping, and the Department of Energy versus the National Institutes of Health. In each case, a compromise solution that favored neither one side nor the other was found.

THE SPECTER OF BIG SCIENCE

Basic research in biology has never before been conducted by large groups of people focused on a single goal. The most important biological discoveries, almost without exception, have been made by small groups working together in a single laboratory. The group typically includes a professor, a few graduate students, perhaps a postdoctoral fellow, and a technician or two.

Biologists tend to be fiercely independent scientists who cherish the freedom to follow their research wherever it leads them. They describe this flexible, unstructured approach as "small science" and speak scornfully of the "big science" projects pursued by physicists and astronomers.

To some, sequencing the genome seems like an immense and monotonous example of big science. It conjures up images of mindless technicians in monstrous labs slowly churning out streams of DNA code: GGA-GAGAACACCAGCTTG on and on, year after year. What could be more boring—or expensive?

Although many scientists and members of the public object to the genome project on these grounds, the sequencing versus mapping debate has actually been resolved. Until sequencing can be done easily and cheaply, the project's first priority will be mapping.

An argument with a bit more sophistication claims that 95 percent of the genome isn't even worth sequencing. Genes—those sections of DNA that code for proteins—account for only 5 percent of the genome, after all. Why bother sequencing the total human genome when all you really want is that crucial 5 percent? The rest of the genome, according to this argument, is simply "junk DNA."

In answer to this argument, consider what Robert K. Moyzis, director of the Center for Human Genome Studies at Los Alamos National Laboratory, has gone on record as saying: "I think the non-protein-coding regions are the most interesting regions of the genome because they are the regions that make it all work. There are many DNA codes other than the protein code, and determining the other codes is probably the most basic scientific justification of the Human Genome Project."[5]

After the Santa Fe workshop, DOE began gearing up for its proposed genome project. In the meantime, the National Research Council and the Office of Technology Assessment began preparing reports on the feasibility of such a project. These reports, eventually published in 1988, were both highly favorable.[6] Slowly—almost reluc-

James Watson discovered the structure of DNA with Francis Crick in 1953. His early support for the Human Genome Project captured the attention of Congress and the public. He became the first director of the project on October 1, 1988.

tantly, it seemed—NIH stepped forward and asked for a place at the table.

As the wheels of politics turned, scientists out in the field continued to plow ahead. With or without a genome project, they were devoted to applying the new DNA analytical methods to problems of molecular biology and medical genetics. Successes were reported. A physical map of the *E. coli* bacteria was finished. Locating the cystic fibrosis gene, narrowed to chromosome 7 the previous year, seemed inevitable. Defenders of the genome initiative pointed to these accomplishments and argued that a genome project would bring order and organization to the hodge-podge of existing efforts.

By 1987, both DOE and NIH requested money from Congress for genome efforts. DOE received $12 million for its early pilot efforts. NIH received $17.2 million to find a director and get its program started.

What NIH needed was someone who could lead a major project. More importantly, it needed someone who could give the project legitimacy in the eyes of Congress and the public. It needed someone with impeccable credentials so that the country could be sure that science, not politics, was ruling the project.

DIRECTOR JAMES WATSON

The NIH administrators didn't have far to look. They wanted James Watson, a vocal supporter of the project, for the job. Watson, although already very busy as director of Cold Spring Harbor Laboratory, accepted the position. He said later that "only once would I have the opportunity to let my scientific life encompass the path from double helix to the 3 billion steps of the human genome."[7] Watson began directing the project on October 1, 1988.

On that same date, DOE and NIH agreed to a Mem-

Bernadine Healy, a physician, was named director of the National Institutes of Health in 1991. She held that position for about one year.

orandum of Understanding that described how the two agencies would coordinate the government's genome project. Together, the two agencies developed the project's first five-year plan. The plan emphasized mapping the genome. It also emphasized developing new sequencing methods before committing the project to large-scale sequencing.

Two of the project's three debates were resolved. As for the question of big versus small science, a compromise for that was worked out too. By encouraging chromosome mapping efforts, the project enabled small teams of scientists, like Joan Overhauser's group, to participate. Most of the small teams would be funded through NIH. At the same time, large, multidisciplinary centers like those at the national labs could work on problems that yielded to such an approach. And NIH's new program of genome centers helped establish large, multidisciplinary groups, such as Emanuel's, at existing academic locations.

The total project cost was estimated at $200 million per year. That's $3 billion for the total effort. It sounds like a lot of money—and it is—yet in comparison to other government science programs it's relatively inexpensive. Of NIH's *yearly* budget of $8 billion, the yearly cost of a fully funded genome project would come to less than 3 percent of that total. By way of comparison, the space station project is billed at $8 billion. The Superconducting Super Collider, the physics project begun in 1990 and canceled by Congress in 1993, was to cost $3 billion.

James Watson successfully steered the Human Genome Project through its first unsteady years. Outrageous at times, candid always, America's best-known biologist converted many of the skeptical into believing that this was, indeed, a project worth doing.

By the fall of 1991, however, Watson and Bernadine Healy, the National Institutes of Health director, seemed headed toward a showdown. Under Healy, NIH had begun filing for patents on DNA fragments discovered by the genome project. Convinced that this would harm the genome effort, Watson vehemently protested Healy's policies to the press and anyone else who would listen. Healy, in turn, summoned him to her office and instructed him to keep his complaints private.

At roughly the same time, Healy began questioning Watson's stock holdings in several biotechnology companies—companies that might, conceivably, benefit from his decisions as genome project director. She said that she feared he might have a conflict of interest.

Just what happened next depends upon whose account you believe. When the dust finally settled, Watson had resigned. He faxed his resignation to Healy on April 10, 1992. In December 1992, Francis S. Collins was named director of NIH's genome effort. He started his new job in April 1993.

When NIH's patent applications were rejected in 1992 and 1993, the agency chose not to appeal the decision.

INTO THE FUTURE

So, who *are* you? You know what your parents, gymnastics coach, and friends have to say. What answer might your genes give? If you are willing to offer some cells for DNA testing and pay for the screening, you could learn plenty about yourself. You could learn if you are a carrier for a number of inherited diseases. If you come from a family in which Huntington's disease or a certain form of breast cancer is common, you could check your susceptibility to those diseases, too.

A decade from now, as the Human Genome Project reaches completion, you should, theoretically, be able to learn even more about your genetic heritage. Scientists are predicting that the genes responsible for cancer, mental illness, and several types of Alzheimer's disease will be revealed.

The best predictions say that such illnesses will be the result of several genes (actually their proteins) acting upon each other. Diet, exercise, and the kind of work you do will likely play a role as well. In other words, the causes of such diseases will be complicated. It will be difficult to tell someone, with certainty, whether he or she will develop the illness.

Some people even think that genes for intelligence, musical talent, and athletic ability will be discovered. This seems like a grossly oversimplified understanding of how genes and proteins work. It also assumes that we all agree on what these abilities are. Is an intelligent person someone who takes tests well? Or someone who can take an engine apart and put it back together? Or someone who can persuade a group of people to adopt a particular plan? Or someone who can memorize poetry quickly? Would a person who inherits "intelligence genes" but is never taught how to read be considered intelligent?

Given all these limitations, would you want to know what was written in your genes? It might seem very tempting. Knowledge, we're often told, is good. And your genes, after all, are part of who you are, just like your physical appearance, nationality, and religion.

Consider the following four scenarios. These situations are hypothetical. Think of them as science fiction. They cannot take place today. Yet each one illustrates the potential—and risk—of this new genetic knowledge. Scenarios like these are being used by both critics and champions of the Human Genome Project to illustrate where the project is leading us. You can find situations like these discussed in newspapers, books, and magazines, on television, and even in school assignments.

SCENE 1

You decide to get a complete and total readout of your DNA. You make an appointment at a genetic testing center. There, a counselor speaks briefly with you before a technician scrapes a few cells from the inside of your mouth. A few weeks later, you receive an enormous printout of the findings. You turn immediately to the report summary and learn that you are a carrier for cystic fibrosis and color blindness. You have many genes linked to heart disease but none of the genes associated with depression or Alzheimer's disease. On and on the summary goes, a smorgasbord of good and bad news.

A single test that could evaluate the 100,000 genes of a single person—why, that's a technical achievement more difficult than sending an astronaut to the moon. As we saw in the Chapter 3, DNA testing is rarely clear-cut. Tests that must search for more than one mutation in a single gene, or for more than one gene, are difficult to administer and interpret. Tests that screen an entire genome might very well be in our future, but we need to appreciate how much technical innovation must still occur before such a thing would be possible.

Nevertheless, this is the kind of scenario that both critics and enthusiasts of the genome project often invoke. From one perspective, the benefits are obvious. Armed with accurate information about your genetic potential, fans say, you will respond rationally. Aware of the possibility of producing children with cystic fibrosis, you insist on prenuptial testing of your future spouse. You begin regular exercise and adopt a low-fat, low-cholesterol diet to counterbalance your heart disease genes. You celebrate the good news regarding depression and Alzheimer's.

But critics of such a scenario are worried. They fear that the discovery of disease genes will drastically change how we think of illness. Up until now, illness was regarded as something that happened to us. Fate or bad luck was blamed when disaster struck. What will it mean to discover that many illnesses may be written in our genes? Will we feel like walking time bombs? Will we look upon sick people with less sympathy, as if their infirmity was a sign of their own weakness?

Other critics of genetic screening worry that the focus on genetic causes will prompt us to stop taking notice of other causes of disease. The Council for Responsible Genetics (CRG), a Boston-based group devoted to preventing genetic discrimination, believes that if our goal is to improve human health, then it makes more sense for us to devote energy and tax money toward improving economic and social conditions than to the Human Genome Project.[8]

Certainly a poor child living in a dilapidated building filled with lead paint and rats is at high risk for health problems, regardless of his or her genes. Smoking causes cancer. Poor pregnant women without adequate nutrition and prenatal health care are at a greater risk for delivering a premature baby, and even the best combination of genes can't rescue an infant born too soon.

The logic of the CRG is indisputable. Yet there is no reason to believe that if the genome project didn't exist, the government would devote more money to social problems and public health. The research would probably continue, but in a slower, less coordinated fashion.

SCENE 2

You are a genetic counselor. This morning you have an appointment with Mr. and Mrs. Garfield. They have four daughters and are desperate to have a son. They've heard about a new technology that combines in vitro *fertilization with DNA testing. In this procedure, four or more eggs are drawn from a woman's ovaries and fertilized with her husband's sperm in the laboratory. Three days later, a single cell from each of the pre-embryos is taken for DNA testing. The pre-embryo with the desired characteristics is then implanted in the woman's uterus. The rest of the pre-embryos are destroyed. The Garfields want to use this technology to pick out a male pre-embryo. What do you say to them?*

This procedure, known as preimplantation genetic diagnosis, already exists. First developed in England, the experimental technique has led to ten or twenty births. So far, scientists have used it only for high-risk couples who wish to avoid passing on genes for a serious genetic illness. But the prospect of making the procedure available to couples for characteristics that don't have anything to do with disease are chilling.[9]

"The use of [DNA testing] for sex selection insults the reasons I went into genetics in the first place," Francis Collins told *Time* in 1994. "Sex is not a disease but a trait!"

Sex selection doesn't have to wait for new lab procedures, of course. In China, where a strong cultural preference for boys exists, amniocentesis and abortion have been used to select for boys. According to reports, about 1.7 million fewer girl babies are born each year than would be expected.[10]

If—and this is a big if—genes for traits like intelligence or athletic ability *are* found, will some parents wish to choose for these traits? Isn't this baby shopping? Suddenly the images of embryos grown in glass tanks as described in Aldous Huxley's *Brave New World* seem awfully close.

This scare scenario, more often than any other, is associated with the Human Genome Project. It's very doubtful, due to the complex interaction of the body's 100,000 proteins, that a single gene or two will be found responsible for something as abstract as "intelligence" or "good looks." Baby shopping for traits such as this can't occur until much, much more is understood about human development. Of more immediate concern, then, is the use of DNA testing for gender and other traits.

(Even selecting against certain diseases raises troubling issues of eugenics, as we saw in the Chapter 6.)

A number of groups—including CRG, the ELSI researchers of the genome project, and government advisory panels—are attempting to anticipate and, when possible, propose solutions to ethical questions such as this. For instance, a report issued by the Institute of Medicine in 1994 clearly stated that "prenatal diagnosis not be used for minor conditions or characteristics." It specifically declared that fetal diagnosis and abortion for sex selection was "a misuse of genetic services . . . and should be discouraged by health professionals."[11] Here, then, is your answer to the second scenario. You should discourage the Garfields from their quest.

SCENE 3

You are a physician specializing in cystic fibrosis gene therapy. You are grateful at the end of every day that you are able to extend the lives of the children and teens who come to you. Now, new advances in gene therapy have progressed so that it is possible to do "germ-line therapy" to eliminate the mutation from the gene pool once and for all. This might seem like a logical extension of your work, you suppose. But something about germ-line therapy makes you uneasy. Maybe there's a reason the cystic fibrosis mutation is so common. You distinctly feel that to eliminate it would be playing God, or at least messing with Mother Nature.

To treat an individual with a serious genetic disease like cystic fibrosis is one thing. The whole point of medicine is to prevent, treat, and cure human illness. But many scientists see the attempt to eradicate the cystic fibrosis mutation from the human race as another thing entirely. They argue that we don't understand human evolution well enough to make such a critical decision.

Germ-line therapy (so called because the egg and sperm are *germ cells*) is still the stuff of science fiction. One way or another, the eggs and sperm of a person (or embryo or pre-embryo) would have to be "treated" to replace the unwanted mutation with a healthy copy of the gene.

Should scientists and doctors be allowed to change the fate of the human race in this manner? In June 1983, as the biotech revolution was swinging into full force, a group of religious leaders held a press conference to address this question. Their answer was eloquent and pointed: "No individual, group of individuals or institutions can legitimately claim the right or authority to make decisions on behalf of the rest of the species alive today or for future generations."[12] A fear of germ-line therapy has led some groups to oppose gene therapy and even DNA testing.

There's another reason to approach germ-line therapy very, very cautiously. What we don't know about human DNA is immense. We don't know what 95 percent of it (the non-protein-coding sequences) is for. We don't know how all the genes work in concert. We also don't know why some genes that are so destructive in pairs—like the cystic fibrosis gene—are so common in certain populations. This prevalence seems to suggest that in a single copy, the genetic mutation is somehow useful. It may play a role in human health that we can't even begin to understand.

We do know that the cystic fibrosis mutation is very old. In an article published in the June 1, 1994, issue of *Nature Genetics*, Spanish scientists reported that the phenylalanine 508 deletion was at least 52,000 years old. It was spread across Europe by a band of modern *Homo sapiens*. The fact that the cell membrane gene variation has existed for so long in the European population suggests that people who have a single copy of it have some survival advantage. Researchers hypothesize that it might prevent death from dehydration and diarrhea by retaining water in the body.

The gene for sickle cell anemia, a serious recessive disorder common to African-Americans, works in a similar fashion. One in twelve African-Americans is a carrier. People who inherit a double dose suffer from a painful, life-threatening anemia. Yet carriers of a single gene are resistant to malaria.

SCENE 4

When you were sixteen, your father died of Huntington's disease, a progressive deterioration of the nervous system that begins at midlife. Knowing that your odds of developing the disease are fifty-fifty, you have your DNA screened. Just as you suspected, you carry the disease gene. Although the DNA test was developed years ago, still no gene therapy exists. You estimate that you have twenty to thirty years of normal life left. You decide to switch careers. But when you start applying for jobs, you discover no one will hire you. Apparently, potential employers have access to the results of your DNA test, and they don't want to invest in someone like you.

In 1995, the U.S. Equal Employment Opportunity Commission ruled that genetic discrimination in employment decisions is illegal. In a similar move, Oregon passed a bill supporting a person's right to keep his or her genetic information private. Other states are considering adopting genetic privacy measures. This legislation is based on a model Genetic Privacy Act drafted with support from the Department of Energy ELSI program.

The 1994 Institute of Medicine report strongly recommended laws protecting confidentiality and preventing discrimination.

In this scenario, the fictional "you" requested testing to gauge the risk for Huntington's disease. Another person, acting just as rationally, may have decided against testing and the risk of receiving devastating news. This illustrates another ethical difficulty: In the absence of a cure, is it better or worse to know that you will one day develop an inherited disease?

While reading these four scenarios, you might have felt very strongly one way or another. Or you might have felt somewhere in the middle. Such is the essence of public debate. Well-informed and well-intentioned people may disagree. Yet the debate itself is important. And so is your contribution to it.

Some people might feel that their side has already lost—or won—because the Human Genome Project exists. Yet the future still remains. It might be like Huxley's *Brave New World*. Or it might be like the utopia described by Marge Piercy in her novel *Woman on the Edge of Time*.

In this futuristic world, set in the year 2137, DNA is completely understood and easily manipulated by the people of Mattapoisett, Massachusetts. They use biotechnology to develop the best crops to eat and fibers to weave. Yet when it comes to people, they have encouraged diversity rather than a master race of perfect human beings. Their village community is filled with a rainbow of races and a wealth of individual talents.

Is this an idealistic view? Yes. Impossible? Maybe. Maybe not.

GLOSSARY

adenine (A)—one of the four nucleotide bases found in DNA.

amino acid—the building blocks of proteins. There are twenty amino acids.

amniocentesis—prenatal test in which amniotic fluid is removed from the pregnant uterus. Fetal cells in the fluid are studied for the presence of chromosomal defects.

centimorgan—unit of measurement used on genetic linkage maps. A centimorgan measures the likelihood that two DNA segments will be separated during crossing over. One centimorgan is equal to a 1 percent chance of separation.

chromosome—rod-shaped structure, consisting of DNA and proteins, that is present in every cell nucleus. Each chromosome contains thousands of genes. Humans have forty-six chromosomes (twenty-three pairs).

crossing over—exchange of DNA between a pair of chromosomes. See *genetic recombination*.

cystic fibrosis—the most common lethal genetic disease in the United States. Patients lack a protein that causes mucus to build up in their lungs.

cytosine (C)—one of the four nucleotide bases found in DNA.

deletion—a missing section of DNA on a chromosome.

DNA (deoxyribonucleic acid)—the molecule that carries genetic heritage. DNA is a two-stranded molecule twisted into a double helix.

dominant gene—gene that makes itself evident when one or more copies are present.

double helix—the term used to describe the spiral-staircase shape of DNA.

enzyme—a type of protein that controls chemical reactions.

exon—protein-coding region of a gene.

gene—a segment of DNA that contains the blueprint for a protein. Through the action of their proteins, genes govern inherited traits like physical appearance. Humans have 50,000 to 100,000 genes.

genetic code—the three-letter code, created by nucleotide triplets, that is read by the living cell as a recipe for building proteins.

genetic linkage map—map that shows the relationship between DNA segments on a chromosome based upon how often the segments are inherited together during crossing over. Distance is measured in centimorgans.

genetic recombination—the exchange of DNA between a pair of chromosomes. See *crossing over.*

germ cell—the cells that produce eggs or sperm.

guanine (G)—one of the four nucleotide bases found in DNA.

healthy carrier—an individual with one copy of a recessive gene.

heredity—the transmission of traits from parent to child via genes.

hormone—a kind of protein that regulates growth, reproduction, and other biological functions.

human genome—a complete set of all the genes of a human.

Human Genome Project—national effort coordinated by the National Institutes of Health (NIH) and the Department of Energy (DOE) to develop the tools and techniques needed to map and sequence all the genes of humans plus five other species.

hybrid cell—a cell that contains pieces of DNA from two or more organisms.

hybridization—the process of joining two complementary DNA strands.

inherited disease—a disease that is the result of DNA mutations.

intron—a non-protein-coding region of a gene.

karyotype—a image of a complete set of sorted pairs of an organism's chromosomes.

linked (genetically linked)—two segments of DNA that are inherited together.

marker (genetic marker)—a segment of DNA whose nucleotide sequence is known. Scientists use markers to determine the approximate position of new DNA segments on a chromosome.

meiosis—process of cell division by which a germ cell produces egg or sperm cells containing half the typical number of chromosomes.

molecular biology—the study of molecules.

mutation—change in DNA that can be inherited.

nucleotide/nucleotide base—the building blocks of DNA: adenine (A), cytosine (C), guanine (G), and thymine (T).

physical map—map that shows the relationship between DNA segments on a chromosome based upon actual distance. Distance is measured in nucleotide base pairs.

polymorphism—differences found among individuals in DNA sequences.

probe—a single-stranded DNA molecule used to detect a complementary strand. It is labeled with either a radioactive isotope or fluorescent dye.

protein—a substance composed of amino acids. Proteins, which include enzymes and hormones, are necessary to maintain the health of all living cells.

recessive gene—gene that makes itself evident when it is present in two copies.

restriction enzyme—protein that seeks out particular DNA sequences and cuts DNA at those sequences.

RFLP (restriction fragment length polymorphism)—variation in the DNA sequence that can be identified with restriction enzymes.

sex chromosomes—chromosome pair 23 in humans: XX in females, XY in males.

sickle cell anemia—a common genetic disorder. A large percentage of patients' red blood cells are misshapen, and therefore cannot transport oxygen throughout the body.

sister chromatids—the two "bodies" created when a chromosome splits during meiosis.

thymine (T)—one of the four nucleotide bases found in DNA.

SOURCE NOTES

1. Sublette, Bill. "The Promise of the Human Genome Project," *University of Virginia Alumni News*. November/ December 1993, p. 27.
2. The profile of Dr. Collins appeared in *Time* on January 17, 1994, p. 54. *The New York Times* also ran a profile of him on November 30, 1993, p. C1.
3. HUGO was described as a kind of United Nations for the human genome in an article titled "HUGO News" that appeared in *Genomics*, pp. 385–386 (1989). The author of the article, Victor A. McKusick, credits Norton Zinder with the phrase.
4. In his book *The Human Genome Project* (New York: Plenum Press, 1991), pp. 237–238, Thomas F. Lee questions the connection between the Department of Energy's growing interest in a genome project and the end of the Cold War. He borrows his argument from Robert

Wright who wrote "Achilles Helix," which appeared in *New Republic* on July 9/16, 1990, p. 24.

5. Cooper, Necia Grant. *Los Alamos Science: The Human Genome Project* (Los Alamos, N.M.: Los Alamos National Laboratory, 1992), p. 81.

6. The two favorable reports on the feasibility of a genome project are: *Mapping and Sequencing the Human Genome* (Washington, D.C., National Academy Press, 1988). This report was prepared by the National Resource Council Committee on Mapping and Sequencing the Human Genome. The second report, *Mapping Our Genes—The Genome Projects; How Big, How Fast?*, was prepared by the Congress of the United States, Office of Technology Assessment, April 1988.

7. Watson, James D., "The Human Genome Project: Past, Present, and Future," *Science*, April 6, 1990, p. 46.

8. The Council for Responsible Genetics has issued a statement titled, "Position Paper on the Human Genome Initiative." The Council's address is 5 Upland Road, Suite 3, Cambridge, Mass. 02140. One of its members, Ruth Hubbard, has written a book on genetic testing and the genome project titled *Exploding the Gene Myth* (New York: Beacon Press, 1993).

9. A description of preimplantation genetic diagnosis for Tay Sachs disease appeared in "Healthy Baby Born After Test for Deadly Gene," *The New York Times*, January 28, 1994.

10. Data for the number of expected female births in China comes from "Planning a Family, Down to Baby's Sex," *The New York Times*, November 11, 1993.

11. The Institute of Medicine report is *Assessing Genetic Risk: Implications for Health and Social Policy* (Washington, D.C.: National Academy Press, 1994).

12. The June 1983 press conference of religious leaders opposed to germ-line testing is described by Robert

Shapiro in his book, *The Human Blueprint* (New York: St. Martin's Press, 1991), pp. 368–369).

CHAPTERS 1 AND 2

The interview with Joan M. Overhauser was conducted on March 22, 1994, in her office. She is a molecular biologist and an assistant professor in the Department of Biochemistry and Molecular Biology at Thomas Jefferson University in Philadelphia, Pennsylvania. She is an author of the following scientific papers:

"Somatic Cell Hybrid Deletion Map of Human Chromosome 18," *Genomics*, 1992, vol. 13, pp. 1–6.
"Molecular Analysis of the 18q– Syndrome—and Correlation with Phenotype," *American Journal of Human Genetics*, 1993, pp. 895–906.
"Report of the First International Workshop on Human Chromosome 18 Mapping," *Cytogenetic Cell Genetics*, 1993, pp. 78–96.

CHAPTER 3

The story of the hunt for the cystic fibrosis gene can be found in three news stories in the scientific journal *Science*. In chronological order:

Roberts, Leslie, "The Race for the Cystic Fibrosis Gene," April 8, 1988, pp. 141–144.
Roberts, Leslie, "Race for Cystic Fibrosis Gene Nears End," April 15, 1988, pp. 282–285.
Marx, Jean L., "The Cystic Fibrosis Gene Is Found," September 1, 1989, pp. 923–925.
The Howard Hughes Medical Institute publication *Blazing a Genetic Trail* (1991) also describes the hunt in a section titled "Stalking a Lethal Gene."

CHAPTER 4

The interview with Larry L. Deaven was conducted on June 9, 1994, by telephone. He is a geneticist and the deputy director of the Human Genome Project at Los Alamos National Laboratory.

The Human Genome Project at Los Alamos National Laboratory is described in great detail in *Los Alamos Science*, No. 20, 1992, a 338-page paperback book published by the lab.

The interview with Beverly S. Emanuel was conducted on March 22, 1994, in her office. She is a molecular biologist and chief of the Division of Human Genetics and Molecular Biology at the Children's Hospital of Philadelphia. She is an author of the following scientific papers:

"Prevalence of 22q11 Microdeletions in DiGeorge and Velocardiofacial Syndromes: Implications for Genetic Counseling and Prenatal Diagnosis," *Journal of Medical Genetics*, 1993, pp. 813–817.

"Microdeletions of chromosomal region 22q11 in patients with congenital conotruncal cardiac defects," *Journal of Medical Genetics*, 1993, p. 807–812.

CHAPTER 6

The interview with David A. Asch was conducted on March 22, 1994, in his office. He is a physician and assistant professor at the University of Pennyslvania School of Medicine in Philadelphia, Pennsylvania. He is an author of the scientific paper: "Reporting the Results of Cystic Fibrosis Screening," *American Journal of Obstetrics and Gynecology*, 1993, pp. 1–6.

CHAPTER 7

The interview with Francis Collins was conducted on March 24, 1994, in his office at the National Institutes of Health in Bethesda, Maryland. He is a physician and molecular biologist, and director of the National Center for Human Genome Research.

CHAPTER 8

The history of the Human Genome Project has been described by several of the main players. See:

Cantor, Charles C., "Orchestrating the Human Genome Project," *Science*, April 6, 1990, pp. 49–51.

DeLisi, Charles, "The Human Genome Project," *American Scientist*, September/October 1988, pp. 488–493.

Watson, James D., "The Human Genome Project: Past, Present, and Future," *Science*, April 6, 1990, p. 46.

The first five-year plan for the Human Genome Project appears in the government booklet, *The U.S. Human Genome Project: The First Five Years FY 1991–1995.* NIH Publication No. 90-1590, April 1990.

A description of James Watson's resignation appears in two articles in *Science*:

Roberts, Leslie, "Why Watson Quit As Project Head," April 17, 1992, pp. 301–302.

Palca, Joseph, "The Genome Project: Life After Watson," May 15, 1992, pp. 956–958.

BIBLIOGRAPHY

Bishop, Jerry E., and M. Waldholz. *Genome.* New York: Simon and Schuster, 1990.

Bodmer, Walter, and Robin McKie. *The Book of Man.* New York: Scribner, 1995.

Cook-Deegan, Robert. *The Gene Wars: Science, Politics, and the Human Genome.* New York: Norton, 1994.

Davis, Joel. *Mapping the Code.* New York: John Wiley & Sons, 1990.

Hubbard, Ruth, and Elijah Wald. *Exploding the Gene Myth.* Boston: Beacon Press, 1993.

Huxley, Aldous. *Brave New World.* New York: HarperCollins, 1989.

Kevles, Daniel J., and Leroy Hood, eds. *The Code of Codes.* Cambridge, Mass.: Harvard University Press, 1992.

Lee, Thomas F. *The Human Genome Project.* New York: Plenum Press, 1991.

Micklos, David A. and Greg A. Freyer. *DNA Science.* Cold Spring Harbor Laboratory

Press and Carolina Biological Supply Company, 1990.

Piercy, Marge. *Woman on the Edge of Time*. New York: Fawcett, 1985.

Shapiro, Robert. *The Human Blueprint*. New York: St. Martin's Press, 1991.

Wills, Christopher. *Exons, Introns, and Talking Genes*. New York: Basic Books, 1991.

Wingerson, Lois. *Mapping Our Genes*. New York: NAL-Dutton, 1991.

GOVERNMENT PUBLICATIONS

Assessing Genetic Risk: Implications for Health and Social Policy. Washington, D.C.: National Academy Press, 1994.

Human Genome Program: Primer on Molecular Genetics. Washington, D.C.: Department of Energy, June 1992.

The Human Genome Project: From Maps to Medicine. Bethesda, Md.: National Institutes of Health, 1995.

Cooper, Necia Grant, ed. *Los Alamos Science: The Human Genome Project*. Los Alamos, N.M.: Los Alamos National Laboratory, 1992.

Mapping and Sequencing the Human Genome. Washington, D.C.: National Resource Council Committee, 1988.

Mapping Our Genes—Genome Projects: How Big, How Fast? Washington, D.C.: Congress of the United States, Office of Technology Assessment, April 1988.

Pines, Maya, ed. *Blazing a Genetic Trail*. Bethesda, Md: Howard Hughes Medical Institute, 1991.

Pines, Maya, ed. *Mapping the Human Genome*. Bethesda, Md.: Howard Hughes Medical Institute, 1987.

Understanding Our Genetic Inheritance, The U.S. Human Genome Project: The First Five Years FY 1991–1995.

Bethesda, Md.: National Institutes of Health, Publication No. 90-1590, April 1990.

NEWSLETTERS

NIH and DOE jointly sponsor a monthly newsletter, *Human Genome News*, which is available to any interested person. Contact: Betty K. Mansfield, Oak Ridge National Laboratory, P.O. Box 2008, Oak Ridge, TN 37831.

The Council for Responsible Genetics has position papers and a monthly newsletter *geneWatch* that are available to any interested person. Contact: Council for Responsible Genetics, 5 Upland Road, Suite 3, Cambridge, MA 02140.

INDEX

Cytosine (C), nucleotide, 17, *18*, 28, *29*
Deafness, and chromosome 18, 14
Deaven, Larry L., 54
  DNA sequencing, 57
  emphasis on math/physics, 62
Deletion syndrome. *See* Chromosomal deletions.
DeLisi, Charles, 89
Department of Energy (DOE), 51–52
  the Alta Summit, 88–91
  center locations, 53
DiGeorge syndrome, 60
Discrimination, 105–106
Disease, and lab animals, 68
DNA (Deoxyribonucleic acid), 8
  hybridization, 30
  and inherited disease, 8
  junk DNA, 93
  as part of chromosomes, *16*, 17
  sequencing, 55–58
  shared between species, 67
  structure of, *16*, 17
  studying, 27–30, *30*
  why study other species?, 63
DOE. *See* Department of Energy.
Dominant gene, 37–38, *39*

Double helix, 17, *29*
Dulbecco, Renato, 91–92
Escherichia coli, 55
  and genetic crossing over, 65
  why study, 64
Edwards syndrome, 21
ELSI (ethical, legal, and social implications), 72–74
  at Los Alamos, 54
Emanuel, Beverly S., 58–62, *59*
Enzymes
  restriction enzyme, *44*, 45
  used to cut and slice DNA, 88
  what they do, 20
Eugenics, 74–77
Ewing's sarcoma, 60
Exons, 48
E.coli, 64
Eye color
  polymorphism, 45
  result of several genes, 20

Facial appearance, 14
Foreign research, 86
Fruit flies, 64–65, *65*
  research locations, 68
  shared DNA, 67
Future, 99–100
  emphasis on math/physics, 62
  scenarios, 100–107